HUMAN ORIGINS

WHAT BONES AND GENOMES TELL US ABOUT OURSELVES

Number Thirteen:
Texas A&M University Anthropology Series
D. Gentry Steele, General Editor

Series Advisory Board
William Irons
Conrad Kottak
James F. O'Connell
Harry J. Shafer
Erik Trinkaus
Michael R. Waters
Patty Jo Watson

HUMAN ORIGINS

WHAT BONES AND GENOMES TELL US ABOUT OURSELVES

Rob DeSalle & Ian Tattersall

Texas A & M University Press
College Station

A PETER N. NÈVRAUMONT BOOK

Printed in Italy.

The paper used in this publication meets the minimum requirements of the American
National Standard for Information Sciences—Permanence of Paper for Printed Library
Materials. ANSI Z39.48-1984.
∞
Catalogue-in-Publication Data record is available from the Library of Congress.
ISBN: 978-1-58544-567-7
        1-58544-567-3

Jacket and book designed by Cathleen Elliott, flyleaf design LLC.

Produced by
Nèvraumont Publishing Company
Bronx, New York

To the Memory of Craig Morris
Exceptional Scientist, Colleague, and Friend

CONTENTS

FOREWORD

Texas A&M University Press' Anthropology Series takes great pleasure in publishing Rob DeSalle and Ian Tattersall's *Human Origins: What Bones and Genomes Tell Us About Ourselves*. Designed as a companion book to complement the American Museum of Natural History's exhibit, the Anne and Bernard Spitzer Hall of Human Origins, the book will provide those who visit the museum with an enriched experience and understanding of the origins of humanity. In addition to those fortunate enough to see the exhibit with book in hand, DeSalle and Tattersall's well-illustrated volume will offer a far larger audience of readers the opportunity for a better understanding of our biological heritage.

Although we have cast our net broadly in search of significant manuscripts for the Anthropology Series, we have focused primarily on books pertaining to our biological nature. We believe that understanding our biological history is of enormous pragmatic importance. Today, we are in a whirlwind dance with technology, the pace is ever increasing, and technology is leading the dance. What makes this so bewildering is that we began this dance such a historically short time ago. Humans have spent well over 95 percent of our biological and cultural history as hunters and gatherers, or simple horticulturalists. We are, in effect, a hunter and gatherer in an industrialist's cloak. Before we can take the first step in regaining control of our world of technology, it is essential we understand our origins, and how we came to be as we are.

While many books have been written on the subject of human origins and many more will be written, the approach DeSalle and Tattersall have taken is innovative and timely. It is integrative, relying on both the traditional fossil record and the most recent studies spawned by mapping of the human genome to understand how we are distinctively human, our origins, and our natural history.

Sharing the Anthropology Series goal of publishing books that appeal to an audience not only within but beyond academic walls, DeSalle and Tattersall have produced a book that is succinct and clearly written. To accomplish this in a fashion that permits readers to understand and have confidence in their conclusions, they have emphasized the *process* of reaching these conclusions and the assumptions upon which these processes rest. Rob DeSalle and Ian Tattersall are internationally recognized scholars in their respective fields of genomics and human paleontology, and together they serve as joint curators of AMNH's Spitzer Hall of Human Origins.

When the opportunity first arose to publish *Human Origins*, as general editor of Texas A&M University's Anthropology Series, I was immediately enthusiastic about the prospect. Upon receiving and reading the manuscript, I found that it exceeded my greatest expectations. We thank the authors and the American Museum of Natural History for the opportunity to be a part of this exciting project.

D. Gentry Steele, Series Editor
College Station, Texas

PREFACE

When it was decided to replace the Hall of Human Biology and Evolution at the American Museum of Natural History, since its opening in 1993 a favorite with the museum's visitors from all over the world, it was hard to suppress a tinge of sadness. At the same time, however, the replacement of this classic exhibition by the Anne and Bernard Spitzer Hall of Human Origins offered an enormous opportunity to break new ground.

From the outset, it was determined that the new hall should depart from tradition and attempt to integrate the fossil record, our habitual source of information about the human past, with the exciting new possibilities offered by the burgeoning science of genomics. Seldom have two branches of science come together so rapidly, productively, and seamlessly. It has been very gratifying not only to have been able to combine the American Museum's traditional strengths in paleontology and anthropology with its rapidly expanding expertise and facilities in the areas of molecular systematics and comparative genomics, but also to have had the opportunity to conceive and execute an exhibition that is without precedent in any institution anywhere in the world.

While this book was conceived as a companion to the Spitzer Hall of Human Origins, it is also a stand-alone work, written for everyone with an interest in the latest discoveries in human evolution and in what new technologies and new kinds of information can tell us about the human past. We hope it will appeal to all who wonder about humanity's place in Nature, and how we got to where we are today, and not simply to those fortunate enough to visit the exhibition in New York.

As joint curators of the Spitzer Hall of Human Origins, we have had the privilege of working with a magnificent team of talented, energetic and dedicated people in the Department of Exhibitions at the American Museum of Natural History. Any large exhibition is an enormously complex undertaking, and we have been nothing short of amazed by the ability of this team of specialists of so many kinds to bring this spectacular project to fruition in an astonishingly short time. We would like to thank everyone individually, but in fear of missing even one of so many names we will simply offer our admiration and most profound thanks to "the team." As the authors of this book, we have also had the opportunity to work with some remarkable publishers, notably Peter N. Nèvraumont, who persuaded us to write it, and Mary Lenn Dixon and her staff at Texas A&M University Press, who believed in it. For the book's handsome design, we are indebted to Cathleen Elliott. Thank you all; it has been a pleasure.

Rob DeSalle and Ian Tattersall
New York City

CHAPTER ONE

THINKING ABOUT OUR ORIGINS

WHEN WE BEGAN OUR CAREERS AS SCIENTISTS, WE NEVER REALIZED THAT PHILOSOPHY WOULD BE SUCH A HUGE PART OF OUR EVERYDAY WORK.

But understanding the role of philosophy in what we do as anthropologists, systematists, and evolutionary biologists is critical not only to how we do our science but also to what our science might mean in the greater scheme of things. This has become even more evident to us during the preparation of the Anne and Bernard Spitzer Hall of Human Origins at the American Museum of Natural History, to which this book is intended to serve as a companion—or as a substitute, in the case of those who are unable to make it to New York.

The major question we approach in this chapter is why we would want to understand our evolutionary past in the first place. In this context, we will discuss the outward uniqueness of our species *Homo sapiens*, as well as our obvious similarities to other organisms on this planet. And we will place both these similarities and uniquenesses in a more general context, philosophical, if you will. For our epistemology (a fancy name for how humans explain knowledge) of human origins depends on the rules we devise for establishing the validity of how we know and perceive those origins.

The "philosophical" approach of this chapter will also allow us to look at this inherently fascinating subject from a historical perspective, starting with ancient—even prehistoric—views and continuing with the many different ways in which the origins both of humans and of the universe that contains them have been explained over the last five or so millennia. We will attempt to make this excursion into the history of attitudes to human origins both temporally and spatially complete, meaning that we will not only examine the changes in how humans have approached their origins through the ages, but also give a multi-cultural perspective by detailing the attitudes toward human origins of a representative selection of cultures across the globe.

The establishment of the historical background will be followed by a closer look at the scientific context of human origins. We strongly believe that by clearly explaining the scientific process, it will become evident which approaches to understanding our origins are scientific and which are not. This subject is an extremely important one in the current intellectual climate, and is an essential one for this book. The last part of this introductory chapter will introduce the human origins "toolbox." By discussing what information we have, and need, to understand the science of human origins, we can introduce the three major components of the human origins toolbox: paleoanthropology, genomics, and evolutionary process.

Human Origins **and Sophistication**

What is it about our human nature that makes us wonder about where we come from? We need only to look upward at the sky or downward to the ground to be puzzled, amazed, and inquisitive about our and their existence. But how did we achieve the ability to wonder about our place in nature and in the universe? One way to understand this important question is to delve into the attitudes of ancient peoples. Some ancient peoples have left written or oral traditions concerning their existence and their place in nature, and some have left more indirect traces. For even if we can't ask humans who lived 25,000 years ago what they might have thought about their existence, we can try to infer this from the material evidence they left behind of their lives. And it turns out that people who lived that long ago were probably very sophisticated in their perception of the world around them and their place in it.

The recently discovered cave paintings at Chauvet, France, and the more famous but younger ones of Lascaux in France and Altamira and El Castillo in Spain, are but a few examples of amazing and extremely beautiful depictions of animals from the period between about 34,000 and 10,000 years ago. [Figure 1] At first glance, these paintings might seem to be straightforward renditions of what their creators saw on the landscape around them. But the selection of animals depicted, the juxtaposition of species, and the almost invariable presence along with them of mysterious signs and symbols make it clear that there was a lot more to these images than simple representation. At one time, for instance, it was thought that these paintings might have been related to the hunt: that the artists believed that the ritual slaying of animals on cave walls—for many animal images appear to be pierced by weapons—would by some sympathetic magic lead to success in hunting them.

But scientists who analyze the diets and lifestyles of these ancient people through examination of food remains at their living sites have determined that they were not regularly eating the animals most frequently seen in the paintings. The cave art is more than a mere menu, and is much more likely the result of sophisticated minds attempting to grapple in a larger sense with their surroundings, perhaps even with their very existence. In an even more immediate sense, we recognize that the art was the product of minds that 21st century humans can relate to. There can be little question that the people who quested in this way 30,000 years ago were able to think not only about a plethora of practical subjects, but about abstract concepts, such as their existence on this planet and their place in nature. In this fundamental sense, these people were modern people.

What is not clear is what their exact thoughts were. Resurrecting entire systems of belief from such indirect evidence is an impossible task. But such exactness is much less of a problem when we come to the written and oral traditions that are recorded for several ancient civilizations. It is amazing that, of the many civilizations that have left or are continuing to leave records of their thought processes, almost all have bequeathed us evidence of their interest in their own origins. Tax records aside, perhaps the most prominent of their concerns seems to have been with what, for lack of a better word, we would call "creation." The almost universal need shown by most civilizations, religions, and philosophical movements of the last five millennia to address this problem of origin extends not only to people, but to Earth itself. The various creation stories and their philosophical

FIGURE 1. Animal images from the "Hall of Bulls" at the cave of Lascaux, in southwestern France. Accompanied by a profusion of inscrutable symbolic signs, these 17,000 year-old images have never been exceeded in their power and expressiveness. Courtesy of Norbert Aujoulat, Centre National de Préhistoire, Ministère de Culture, France.

ramifications are deeply affected by the underlying thought processes. Even a brief examination of the philosophical issues associated with creation myths will make it clear that the fundamental structures of the creation stories used to explain the natural world around us are very different from the structures of scientific accounts. Of course, both religious and scientific approaches to human and planetary origins involve protagonists and events. But scientific approaches have a unique philosophical framework all to themselves, and an understanding of this framework is critical to how we, as scientists, approach our own origins and the origins of the natural world around us.

Creation Myths and the Philosophy of Existence

Creation myths from different cultures are unique in their details, but the thought process that lies behind them is pretty universal, and is based on an apparently innate desire among people to know their roots and heritages and places in nature. [Figure 2] This is why the same general ideas in creation myths are repeated over and over in very different cultures.

While it is difficult to categorize these myths, there are some commonalities among them. For instance, the Hmong, Maya, Navajo, and Norse all suggest that man arose from some plant matter (pumpkins, maize dough, white and yellow corn, and logs respectively). But man can also come from other substances, such as clay (Creek and Judeo Christian accounts), dirt (Inkan accounts), and eggs (Egyptian and Surat Shabda Yoga accounts). In other recountings, individual gods split in two to form Earth and water, as in ancient Babylonian mythology and various ancient Chinese accounts of creation. And in still others, humans are the offspring or creation of the gods (Maori myths, Zoroastrianism, ancient Greek, Inca, Christianity, Islam, and others). Some approaches

FIGURE 2. Two creation myths. Below: The creation of the cosmos and the separation of heaven and Earth are shown on the painted sarcophagus of Butehamun, scribe, in the necropolis of Thebes. The star-covered body of the sky-goddess is being lifted up by the air-god. 21st Dynasty. Turin, Museo Egizio. Above: Detail from the Sistine Chapel ceiling by Michelangelo.

to creation, such as those of conservative Christianity, Judaism, and Islam, have emphatic ideas embedded in their written records (the Bible and the Quran) that implicate a creator in the origin both of man and the universe.

Even refinements of these strongly phrased and rigidly posed accounts, such as intelligent design is to conservative Christianity, implicate a creator as the ultimate cause of the origin of complex things. Other accounts of creation skirt the issue altogether—such as those of Buddhism and Taoism. Buddhism suggests that questions about creation are not connected with the goal of life, so that ideas about creation are not a part of "holy life." Taoism suggests that because we don't know the real nature of creation "we can only call it the Way." Oral accounts of human origins, such as those of some native American groups (e.g., Iroquois, Inuit, and Huron) involve women, while others, such as those of the Yoruba in western Africa, leave women out completely.

While the conservative Christian account of creation is often claimed to be at odds with more natural explanations for the existence of the universe in general and humans in particular, many other accounts from other religions are not in conflict at all—and almost all of them work at the level of metaphor. We have already mentioned Buddhism and Taoism as religions that avoid the issue of creation and hence have no conflict with ideas of evolution; but Hinduism, Zen, Japanese mythology, and many specific branches of Christianity also see no conflict between their doctrines and natural explanations for the origin of universe and man. For instance, the Catholic Church made the following statement on Oct. 23, 1996: "Fresh knowledge leads to recognition of the theory of evolution as more than just a hypothesis." The statement clarified a longstanding attitude of the Catholic Church that a natural explanation of the universe, such as evolutionary theory, does not conflict with Catholic doctrine. The rider to this statement, however, makes it plain that accounts of evolution are only okay as long as the ultimate creator of the human soul is God. And it appears that as leaders change in religions, so might attitudes toward science. This notion is exemplified in a recent *New York Times* editorial by Christoph Schönborn, the Roman Catholic cardinal archbishop of Vienna. In this editorial, Schönborn shifts the attitude toward evolution from one of non-competition to one of direct confrontation:

> Evolution in the sense of common ancestry might be true, but evolution in the neo-Darwinian sense—an unguided, unplanned process of random variation and natural selection—is not. Any system of thought that denies or seeks to explain away the overwhelming evidence for design in biology is ideology, not science.

The critical word in the quote above is "design" about which we will have more to say later in this chapter.

The breadth of these accounts of the beginning of the universe and the origin of man illustrates the importance to the human mind of understanding the origin of the natural world around us and our own place in it. But there are different approaches to doing this. The religious and mythical accounts are united by one basic approach—faith. In order for these accounts to mean anything with respect to the natural world, one must have faith in the accounts, faith in a creator, or faith in an initial cause for the creation of the universe. No amount of evidence (short of the physical appearance of a creator) can help to evaluate the accuracy of such explanations of the origins of the universe, Earth, and humans.

Animal Parts and Sailing Beagles

While all creation myths and stories have traditionally satisfied humans' need for explanation of their surroundings, more precise descriptions of nature have been developed since the birth of modern science. Over the past two centuries or so, a completely new way of thinking about nature has emerged; but ancient science is another matter, and it makes an interesting halfway house in ways of looking at the world. Aristotle was the first writer to detail the various organisms in nature and to try to make some sense of the lavish diversity of the natural world. In his *Parts of Animals*, he described some 500 different organisms, with great precision and some speculation. Imagine being the first person in the history of humans to realize and write the following:

> Of animals, some resemble one another in all their parts, while others have parts wherein they differ. Sometimes the parts are identical in form or species, as, for instance, one man's nose or eye resembles another man's nose or eye, flesh flesh, and bone bone; and in like manner with a horse, and with all other animals which we reckon to be of one and the same species. *Parts of Animals* Book I; Section 1.

This passage demonstrates an acute realization of the form and substance of animals and the suggestion that parts of one animal will resemble parts of other animals. But Aristotle could not quite get away from considering animals as fixed creations, and indeed today his ideas about the generation of animals seem quite bizarre to us:

> Of these insects the flea is generated out of the slightest amount of putrefying matter; for wherever there is any dry excrement, a flea is sure to be found. Bugs are generated from the moisture of living animals, as it dries up outside their bodies. Lice are generated out of the flesh of animals. *Parts of Animals* Book V; Section 31.

In Aristotle's view animals remained fixed, unchanging entities in nature. They were generated, not evolved. His approach was to describe the type of organisms, to extract the average characteristics of organisms. This forced him to think in a very narrow sense about nature, and his great contribution to our understanding of it lay in his acute observational and descriptive abilities. But at the same time, this meant that Aristotle was unable to recognize the importance of biological variation. While in Books VIII and IX of the *Parts of Animals* he did recognize that variation existed—as is evident in his detailed descriptions of varieties between and within species—he simply subdivided this variation further, and described the resulting varieties as types.

The Christian tradition of observing the universe retained the creation myth as its basis and reaffirmed Aristotle's type-based thinking, which thus directly influenced the way nature was viewed for nearly two millennia. Other societies, cultures, and religions also imported a semblance of the creation myth into their observations of the natural world; and indeed, an element of storytelling is implicit in almost any description of nature. Many folk descriptions of nature in general and of organisms in particular are derived from creation myths. But the main common thread of Aristotelian and Christian views was the idea that species were fixed. The idea that species were unchanging entities fits very nicely with the Christian creationist account of the origin of life and of humans, in which, the Creator had made all of the species on the planet just as they exist today.

In this brief discussion of evolutionary thinking we have succumbed to the temptation to jump immediately to Charles Darwin and the influence of the observations he made during the voyage of the *Beagle*. [Figure 3] For aside from some wonderful developments in early Islamic Arabian science, the rumblings of a great leap forward in science during the Renaissance, the opening of the mind to studies of nature, and the rise of anatomical studies in Europe just prior to Darwin's work, the period in between Aristotle and Darwin (over 2,000 years) is remarkably barren of any new thinking about origins. The Dark Ages were truly dark with respect to the human desire to understand the place of humans in Nature; and the Renaissance focused mostly on the physical sciences, as did Arabian science. Most of the anatomical studies of the late 1700s and the early 1800s were shackled by religious beliefs about the origins of life, species and man. It took an immense stride in thinking to break away from the two millennia-old view of our origins.

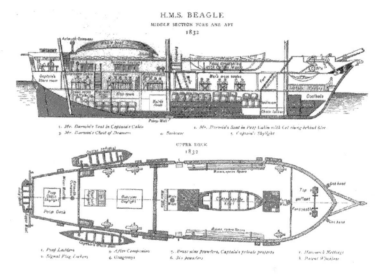

FIGURE 3. Cross sectional diagram of the HMS *Beagle*, the ship that Darwin sailed on to South America and the Galapagos.

Darwin's contribution to our modern scientific perspective was his articulation and defense of the idea that evolutionary processes, and particularly natural selection, have molded the malleable natural world around us. His ideas about nature began to form when he was a young unpaid naturalist on the global circumnavigation of the HMS *Beagle* from 1831 to 1836; and they had, indeed, matured considerably by the time he left the vessel. But then, for two agonizing decades, he kept his ideas to himself. He was forced to make them public, and to display the reasoning that had formed them, by the emergence of similar thinking from his younger contemporary Alfred Russel Wallace and in 1859 Darwin published his world-changing perspective in his great work *On the Origin of Species by Means of Natural Selection*.

This massive volume is rightly regarded as among the most seminal works in the entire history of science. What did Darwin do right to merit his amazingly important place in human thought? Of the many things he achieved in the *Origin of Species*, three stand out. The first is his presentation of overwhelming evidence in favor of the process of evolution as the explanation for the pattern evident in the natural world. Darwin called his treatise "one long argument" and his friend and supporter Thomas Huxley called it "a long chain of arguments," for the existence of evolution. His second major achievement was to present a possible mechanism for evolution. This he called "natural selection," and we will have much more to say about it later. His third achievement was to add a completely new approach to observing things in nature. Remember that Aristotle looked to the mean, or the average, type when he wanted to describe things. If variation existed, he either melded it into an average or type description, or he called the variation a new thing entirely.

In contrast, Darwin turned this way of thinking upside down and began to believe that the variation he saw in nature was real, while the mean or type was an abstraction. What this allowed Darwin to do was to begin thinking about how nature dealt with variation, and to recognize that species were *not* fixed. Once he had recognized this, the floodgates opened, and his revolutionary perspective became the pervasive and persuasive way of explaining nature. In particular, Darwin saw that natural selection acted upon the variation he observed in natural populations, and that he could use this in explaining how evolution happened. It was because he thought in this new way that he could see that species were not fixed. In fact, the only figure in *On the Origin of Species* is a genealogy, or tree, that shows the lack of fixity of species and the fluidity of transitions of forms from generation to generation. We will have much to say in subsequent chapters about evolutionary trees.

While totally committed to evolution as an explanation of natural diversity, Darwin personally found the results of his novel thought processes excruciating. Later, he admitted that when it dawned on him that evolution was indeed a possible way of explaining nature, that species were not fixed, and that natural selection was a possible mechanism to explain that fact, it felt like "confessing a murder." For the past 150 years, scientists have taken Darwin's astonishing insights to heart, and by dint of a lot of hard work and reasoning, have molded them into a diverse but coherent branch of biology.

Eureka! The Philosophy of Science

What exactly can we say is real or true in the universe? Is knowing something different from explaining something? How do we learn things about our biological past? These are all questions that affect our philosophical approaches to our origins, and that bear directly on our ability to assess our scientific approaches to understanding our universe and our origins. The postmodern notion of the natural world, for example, suggests that knowing things for certain is not possible. What this means in a scientific context is that when we study history, especially natural history, there is no assurance that we will discover the truth — or even that there is any need to try. The truth simply is unknowable with respect to natural history, where we lack the ability to go back in time and observe the events anew. Some philosophers of science have even extended this notion of unknowable truth to laboratory sciences, such as chemistry and physics. But if truth in the sciences is not attainable, what, then, is it that scientists do, especially in the historical sciences, those most relevant to understanding our origins?

This question is directly relevant to one of those just asked: Is knowing something different from explaining something? The answer to this question is yes, definitely; and since the postmodern view of the world tells us that we cannot know things for certain, then we are left with explaining things. Explaining is our substitute for knowing, and science is about finding the best explanations for natural phenomena.

One of the best-known philosophers of science in the 20th century was Karl Popper, who was preoccupied with how science worked. His conclusions are highly relevant to the study of the origins of the universe and of humans. Popper proposed that there is very little that we can know for sure. He suggested that the best explanations for things in the natural world come from very severe criticism, or testing, of ideas or hypotheses. Popper also recognized that because we cannot know the truth, the next best thing is to be able to show that certain ideas or hypotheses are false—for verification, the search for proof, is a shaky basis for science.

The classical example of this situation is the white swan statement. We can make the statement that "all swans are white." We can superficially confirm or verify this hypothesis by finding a few white swans and then just quit looking. But to prove or confirm the statement completely, we would need to find all the swans on the planet and verify that they are all white, an impossible task. When one adds the dimension of time to the problem (if you say "all swans," then that should also include swans in the past and the future, shouldn't it?), the verification of the statement becomes impossible, unless one has a time machine.

What *is* possible, though, is to falsify, or reject, the statement that "all swans are white." The mechanism of falsifying this statement exists simply in finding a swan that is not white. [Figure 4] Such do indeed exist, in the form of *Cygnus atratus*, the Australian black version of swans. The existence of the Australian black swan then negates or rejects the initial statement that "all swans are white." The point here is that while you can often verify statements on a superficial level, you can never do it universally; but when a statement is falsified once, it drops out of contention. There is a huge difference, then, between verifying a statement and testing it to attempt to falsify it.

In Popper's perspective, then, statements that are constructed so as to be verified are unscientific. Only statements that are framed so that they can be tested or falsified are scientific; and any statement that cannot be refuted is not scientific. Science is not an authoritarian system, and contrary to popular belief, irrefutability is not a desirable characteristic of scientific theory. Indeed, the only advances we make in science come at the expense of refuting statements that have been made. Every day in science, perfectly good

FIGURE 4. *Cygnus atratus*, the Australian swan that falsifies the assertion that "all swans are white." Courtesy Wikipedia.

statements or hypotheses are put to the test, and refuted. And it is through the falsification of such hypotheses that we refine our knowledge of the natural world around us.

Through knowing what does not exist or what is not real (and those things or statements that are falsified help us to decide this), we are able to describe the natural world more accurately. In fact, this approach of falsifying things allows us to come up with the best available explanations for the things we see around us. We won't know if our explanations are true, but we can be assured that they are the best we can do, at least for the moment, to explain the natural phenomena around us. What this also means is that many ideas in science are highly provisional and prone to rejection by the process that Popper described.

Why make such a big deal out of falsifiability and testability? Because, unlike the creation myths described in the previous section of this book, the basics of evolutionary theory are testable. Even the hard-to-please Popper, after much thought, came to the conclusion that evolutionary theory offered "a possible framework for testing scientific hypotheses." In other words, evolutionary questions can be posed in the form of statements that are susceptible to falsification. And indeed, over the past century, thousands of hypotheses relating to the evolutionary process have been tested, and many rejected. This process, of testing and rejecting hypotheses about the patterns of relationship among the organisms on this planet has led to the recognition of common ancestries as the best explanation for the pattern of life we see around us (sets within sets).

Actually, it turns out that evolution is the *only* approach to explaining our natural world that predicts the pattern we actually find out there. The other possible alternative currently proposed to explain what we see around us is special creation (or, in its most recent disguise, "intelligent design"). [Figure 5] Creation by a supernatural being would imply only that the world is the way it is because that supernatural being decided to make it that way. It makes no predictions about the pattern we actually see—whereas evolution does. The result is that the creationist explanation for our natural world is untestable. To accept creation as an explanation requires faith. And although faith is quite legitimately untestable, it isn't science.

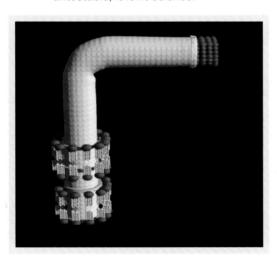

FIGURE 5. A flagellum—the symbol of ID. according to Intelligent Design proponents, this structure is supposed to be irreducibly complex.

In an amusing critique of the "science" in intelligent design the writers at the humor newspaper *The Onion* offered this headline: "Evangelical Scientists Refute Gravity With New 'Intelligent Falling' Theory." [Figure 6] This cleverly written satire points out the problems with "faith-based science." The fictional conservative "intelligent falling" scientist quoted in the article makes the following statement: "Things fall not because they are acted upon by some gravitational force, but because a higher intelligence, 'God' if you will, is pushing them down." An interesting statement, but entirely untestable.

In the context of "intelligent falling," no one in his right mind would suggest that the concept of gravity is not a natural force, because of the many scientific hypotheses that have been tested and falsified about gravity since Newton's time. The series of testing experiments on this question over the past several centuries has accumulated to the point that the only available explanation for gravity is a natural one. Note that over the past centuries several ideas about, and formulations of, gravity were in vogue before ultimately being rejected after scientists tested them and found them wanting. The provisional nature of the current formulation of gravity has at no time in its development jeopardized its natural explanation. "Intelligent falling," or any other faith-based explanation for gravity would, we know, be untestable and hence unscientific. Likewise for the faith-based explanations for the origin of the natural world around us, and for the origin of our species in particular.

Evangelical Scientists Refute Gravity With New 'Intelligent Falling' Theory

August 17, 2005 | ISSUE 41 • 33

KANSAS CITY, KS—As the debate over the teaching of evolution in public schools continues, a new controversy over the science curriculum arose Monday in this embattled Midwestern state. Scientists from the Evangelical Center For Faith-Based Reasoning are now asserting that the long-held "theory of gravity" is flawed, and they have responded to it with a new theory of Intelligent Falling.

ENLARGE IMAGE

"Things fall not because they are acted upon by some gravitational force, but because a higher intelligence, 'God' if you will, is pushing them down," said Gabriel Burdett, who holds degrees in education, applied Scripture, and physics from Oral Roberts University.

Rev. Gabriel Burdett explains Intelligent Falling.

FIGURE 6. Intelligent Falling: A fictional headline and news story from the humor newspaper *The Onion*.

A Toolbox for Human Origins

Throughout this book we will delve into the scientific discoveries, experiments, hypothesis tests, and theories about human origins. As with any project, we need tools to explain these subjects well—a toolbox, so to speak, for human origins. The toolbox for human origins 40 years ago consisted of three major tools—paleoanthropology, genetics, and evolutionary theory. The big difference in the toolbox of today is that all three of these tools that already existed in the 1960s and 1970s have been greatly refined and expanded, rather as the toolboxes our fathers used around the house look very primitive, compared with the ones we use now around ours. The old screwdriver is present in a power version, the handsaw is replaced by an electric programmable round saw, and the spirit level has been evicted by a laser-based leveling device.

In our human origins toolbox, the paleoanthropology of 30 years ago has been augmented by 30 year's worth of new discoveries and the invention of more elegant ways of examining the fossil record. The single-gene genetics of 30 years ago is replaced by genomics, in which we are able to consider entire genomes, i.e., all an organism's DNA (deoxyribonucleic acid), or very large portions of them, at the same time. And the older evolutionary theory has been refined, tested, and revised to yield a more coherent body of knowledge with new analytical tools.

THE PALEOANTHROPOLOGICAL TOOLS

The study of human fossils goes back over 150 years, but paleoanthropology as we know it today is only as old as our fathers' toolboxes—less than half a century. Before the middle of the 20th century, most of those who studied human fossils were human anatomists, but in the 1960s, Louis and Mary Leakey assembled a group of specialists to help them study not only the human fossils they had found at Tanzania's Olduvai Gorge, but their age and environmental and archaeological contexts as well.

Thus was born paleoanthropology as an essentially collaborative enterprise, with specialists of many kinds contributing to the understanding of the fossils that make up the documentary core of human evolutionary history. Nowadays it is rare for an expedition aimed at discovering human fossils to go into the field without a whole array of members with widely differing expertise. Indeed, paleoanthropologists these days are frequently part-time, in the sense that their day jobs may well be as stratigraphers, anatomists, isotopic chemists, geochronologists, taphonomists, or many other professions. This means that while many paleoanthropologists are highly specialized, the toolbox available to paleoanthropology as a whole is amazingly broad.

Paleoanthropologists look for and collect human fossils, but they also look at how the remains of dead animals may have been preserved from destruction to become fossils. They look at the rocks that contain the fossils, and they study what those rocks can tell us about the environments in which our ancient precursors lived. They use a variety of ingenious techniques for dating those rocks, and thus for knowing when the fossil remains they collect were living, breathing animals on the landscape. They study the fossils themselves, and they do this from a whole host of perspectives. Some paleoanthropologists are most interested in identifying the species the fossils represent, so that they can proceed to figure out what parts those species played in the evolutionary story, and which other animals they were related to. Others specialize in studying how extinct human species functioned in life, what they ate, and even how they may have behaved in the social setting.

One unique feature of humans is that, since the invention of stone tools some 2.5 million years ago, they have left us a material record of their existence that extends far beyond fossils. This is the province of the archaeologists. As we'll see, the record of early human activities is very indirect and incomplete, but it has given us an amazing insight into the behavioral history of our predecessors. Like the more biological and geological areas of paleoanthropology, Stone Age archaeology itself is becoming increasingly subdivided into specialties.

But paleoanthropology doesn't stop there. It borrows from its sister sciences more liberally than any other specialty. Neurologists, demographers, physiologists, ethnographers, primatologists, ethnologists, philosophers, and a host of other kinds of scientists and scholars find their work freely appropriated by paleoanthropologists, who, after all, seek to understand both humanity and the background from which we emerged in all of their many dimensions. Even parasitologists have lately become involved, for the good reason that virtually any aspect of biology finds an expression of some kind in the species *Homo sapiens* and its precursors. This makes paleoanthropology both intrinsically among the most exciting of the sciences, and perhaps the trickiest to characterize. We'll look more closely at the sheer diversity of paleoanthropology in the next chapter.

THE GENOMICS TOOLS

If one wants to follow the history of modern humans, or of any living organism for that matter, genes are a valid way to go. As we will discuss in later chapters, genes are the hereditary units of organisms. Genes are made up of a substance called deoxyribonucleic acid, or DNA. Because of DNA's unique structure it is an ideal molecule for carrying information. And because DNA can be copied readily and, for the most part, faithfully, it is also well-suited for transmitting information from one generation to the next.

Each species has a more or less set number of genes that are arranged in groups on large structures called chromosomes. The Indian muntjac, (a tiny barking deer), has the fewest chromosomes of any mammal (three chromosomes that pair to make six in most cells of the body), and all of its genes (probably more than 25,000 of them) are packed onto these three chromosomes. Humans have 23 pairs of chromosomes into which all of our 25,000 genes are packed. Almost every cell in our bodies has these 23 chromosome pairs in it, so that our bodies are a collection of millions and millions of sets of 23 pairs of chromosomes. All of these 23 chromosome pairs together make up what we call the human genome. Understanding how these 23 pairs of chromosomes are inherited can help geneticists and human biologists understand the history of human populations and of human origins.

Prior to the 1970s, genetics in the study of human origins used mostly blood-group analysis. Everyone today should know his or her ABO blood type, because blood type is critical for the success of blood transfusions. The ABO blood-type system is controlled by a gene that determines the structure of blood cell surface molecules that can vary from human to human. Researchers in human origins would collect blood samples from a wide variety of people, determine the various blood types, and use the blood-type frequencies as a proxy for understanding the underlying genetic changes. While this information was important, and molded some of the first hypotheses tested about human genetics, the technique could only obtain information on a small portion of the human genome – either one gene or a very few genes.

Later, a way of looking at proteins not associated with red blood cells was developed, but even these protein techniques could only process a few or at most tens of genes for examination. Things changed drastically with the development of modern molecular biology, and of high throughput DNA-sequencing techniques that allow us to characterize the DNA itself. The genomic revolution began in the 1980s, when a group of scientists began suggesting that the entire human genome, that is, all of the genetic information in the chromosomes of a human, could be "read" using biochemical methods that we will describe later. In the early 1990s, the technological advances that were needed to automate the deciphering of the bits of information in a genome (called bases or nucleotides) were invented. The first genome of a living organism was sequenced in 1996. The organism was *Haemophilus influenzae* or plain old "H flu." The completion of

H flu was followed by completion of the baker's yeast genome, followed by a nematode (a worm) genome and the fruit fly genome. In 2000, the National Human Genome Research Institute at the National Institutes of Health (NIH), in coordination with a company called Celera Genomics, announced the finishing of the first draft of the human genome. Nearly all of the 3 billion bits of information in the human genome had simultaneously been deciphered by both NIH and Celera in this first-draft preparation. Shortly after this announcement, the mouse genome and some plant genomes were finished, and in the last half of 2005, scientists at NIH announced the completion of the second full primate genome—the chimpanzee's—a subject that we will touch upon later.

While the sequencing of whole genomes opens up enormous potential for understanding the workings of the human genome, the techniques that were developed to sequence the human genome can be used to examine large numbers of humans for large numbers of genes. These new techniques have revolutionized the ways human genes are interpreted in the context of human origins and human movement.

THE EVOLUTION TOOLS

Evolutionary biology did not become a fully fledged discipline until scientists figured out a way to meld the ideas of descent with modification and natural selection with the mechanisms of genetic transmission. The basic rules of genetics were published by the Czech cleric Gregor Mendel in 1866, but it was not until after they were rediscovered in 1900 that the melding of evolution and genetics began. Experimental, theoretical, and mathematical approaches to evolution followed, and ultimately a notion of evolutionary biology called "the Modern Synthesis" came about, wherein population genetics and evolution were integrated, essentially by reducing evolutionary phenomena to generation-by-generation changes in gene frequencies within populations, under the guiding hand of natural selection.

Along the way, though, evolutionary biologists discovered that in addition to natural selection, several other factors could influence changes over time—among them small population sizes, large amounts of mutation, inbreeding, and large amounts of migration.

Of these four new influences one—small populations—is widely considered one of the most important in evolutionary biology in general, and in human origins in particular. This is because only in small populations can random chance radically alter the frequency of genes. To understand this, try flipping a coin 200 times straight and counting the numbers of heads and tails. When you are finished, your totals of heads and tails will almost always be about 100 heads and 100 tails. Now try flipping the coin four times. Chances are pretty high that you might flip heads all four times. If you don't get four heads (or four tails) the first time, try it again and again; we guarantee that you won't have to do this many times before you get all heads or all tails.

Now think back to genes in human populations. What might happen if a population fell to very low numbers due to disease or a natural disaster? The answer is that the same randomness of the results that you find when you only flip your coin a few times takes over. The phenomenon with the coin is called "sampling error"; in humans (and other organisms) it is called "bottlenecking." Another way to make human populations prone to sampling error is to separate off a very small part of the population through emigration. In this case, the people who move are called "founders," and the skewing of gene frequencies in the next generation produced by those founders is called the "founder effect."

Not only has our capacity to understand how genes behave in populations advanced, but the tools available for the reconstruction of historical events have also expanded. Think back to the figure we mentioned in Darwin's *Origin of Species*. It was a tree, and through the development of genomics we now have available tree-building techniques that can analyze DNA sequence information. These are particularly useful because some DNA sequences evolve faster than outward appearances do. This means that variation at the DNA sequence level is almost always higher than the variation at the morphological level, which in turn allows us insights that we could not get purely from morphology. New techniques for dealing with the onslaught of DNA sequence information in evolution and tree building, as well as the development of large-scale computational tools for building trees, have also become a large part of the basic toolbox of evolutionary biology.

Along with an understanding of what is going on in populations at the level of the genes, and of how new features are incorporated into genomes and gene pools, huge advances have also been made in the way we view larger evolutionary patterns. In reducing the evolutionary process largely to a matter of slowly changing gene frequencies under natural selection, the Modern Synthesis produced an expectation that evolutionary change as seen in the fossil record would be a gradual affair, with species changing slowly over time as they adapted to new conditions or became fine-tuned to their old ones. Any discontinuities seen in the fossil record were attributed to that record's famous "gaps"— and there's certainly no argument with the fact that known fossils represent only the tiniest fraction of all organisms that ever lived.

Yet, as we'll explain, we can see now that there is much more to evolution than that: Evolutionary processes operate at many levels, all the way up from the genes and individual organisms to local populations, species, and even entire ecological communities. What's more, evolutionary change seems most often not to be continuous, but to happen in spurts; and the mechanisms underlying the appearance of new species and new morphologies are not the same. Indeed, it seems to be inaccurate to speak of *the* evolutionary process; in the long history of life many different processes have intervened, all of which have left their mark on its evolutionary outcome.

Why All of This is Important

As we were writing this chapter, yet another challenge to the teaching of evolution and natural origins in United States classrooms loomed just a couple hundred miles from our offices in New York City. Ironically, on the 80th anniversary of the Scopes trial, at which the renowned lawyer Clarence Darrow and the populist leader William Jennings Bryan argued the evolutionist schoolteacher John Scopes' fate in a Dayton, Tennessee, courtroom, the judicial system in Dover, Pennsylvania, faced a similar challenge. But this time, it came in the guise of a newly coiffed, dressed-up and manicured name for creationism: "intelligent design." In Dover, the U.S. District Court for the Middle District of Pennsylvania was hearing a challenge to the way evolution is taught in the town's public high school. A lawsuit (Kitzmiller *et al. vs.* Dover Area School District *et al.*), brought by a group of parents, contested a policy implemented by the city of Dover's school board requiring that a disclaimer be read in science classrooms calling attention to alleged failings in evolutionary theory and suggesting that alternatives be examined in the form of a so-called "new" way of looking at the problem of origins called "intelligent design." [Figure 7] By a 6-3 vote, the Dover school board approved the following resolution:

> Students will be made aware of gaps/problems in Darwin's theory and of other theories of evolution including, but not limited to, intelligent design.

At the time of the imposition of the disclaimer, the Dover school board was led by conservative Christian members of the community, and the parents who filed the suit were angry that what they considered a nonscientific idea (intelligent design), was to be taught as science in their classrooms. Here is what the disclaimer said:

> The Pennsylvania Academic Standards require students to learn about Darwin's Theory of Evolution and eventually to take a standardized test of which evolution is a part. Because Darwin's Theory is a theory, it continues to be tested as new evidence is discovered. The Theory is not a fact. Gaps in the Theory exist for which there is no evidence. A theory is defined as a well-tested explanation that unifies a broad range of observations. Intelligent Design is an explanation of the origin of life that differs from Darwin's view. The reference book, Of Pandas and People, is available for students who might be interested in gaining an understanding of what Intelligent Design actually involves. With respect to any theory, students are encouraged to keep an open mind. The school leaves the discussion of the Origins of Life to individual students and their families. As a Standards-driven district, class instruction focuses upon preparing students to achieve proficiency on Standards-based assessments.

For six weeks, testimony from many experts on both sides ensued. The testimony focused on whether intelligent design (ID) is, in fact, scientific. Two aspects of this matter were examined in the trial. First was the question of whether intelligent design meets the criteria to be called science. The second and more practical question was whether publications on intelligent design are properly peer-reviewed, as is routine in science. While we were finishing this chapter, the decision came down from Judge John E. Jones III. His 139-page opinion is stunning enough for us to look at it in some detail.

Kenneth Miller of Brown University was the lead witness for the prosecution. Miller has written several biology textbooks, and he testified at length concerning the lack of scientific attributes in intelligent design. Robert Behe of Lehigh University in Pennsylvania was the star witness for intelligent design. Behe has applied the concept of irreducible complexity to only a few select systems: (1) the bacterial flagellum; (2) the blood-clotting cascade; and (3) the immune system. Behe's testimony was soundly refuted by Miller, as summarized in the following quote from the judge's decision:

> Contrary to Professor Behe's assertions with respect to these few biochemical systems among the myriad existing in nature, however, Dr. Miller presented evidence, based upon peer-reviewed studies, that they are not in fact irreducibly complex.

In fact, there isn't a single peer-reviewed scientific article on intelligent design. More to the point, the judge addressed the scientific approach in intelligent design:

Case 4:04-cv-02688-JEJ Document 342 Filed 12/20/2005 Page 1 of 139

IN THE UNITED STATES DISTRICT COURT
FOR THE MIDDLE DISTRICT OF PENNSYLVANIA

TAMMY KITZMILLER, et al.	:	Case No. 04cv2688
	:	
Plaintiffs	:	Judge Jones
	:	
v.	:	
	:	
DOVER AREA SCHOOL DISTRICT, et al.,	:	
	:	
Defendants.	:	

MEMORANDUM OPINION

December 20, 2005

INTRODUCTION:

On October 18, 2004, the Defendant Dover Area School Board of Directors passed by a 6-3 vote the following resolution:

> Students will be made aware of gaps/problems in Darwin's theory and of other theories of evolution including, but not limited to, intelligent design. Note: Origins of Life is not taught.

On November 19, 2004, the Defendant Dover Area School District announced by press release that, commencing in January 2005, teachers would be required to read the following statement to students in the ninth grade biology class at Dover High School:

> The Pennsylvania Academic Standards require students to learn about Darwin's Theory of Evolution and

1

FIGURE 7. The front page of the 139 page landmark Dover "monkey trial" decision made in 2006 by Judge John Jones

Accordingly, the purported positive argument for ID does not satisfy the ground rules of science which require testable hypotheses based upon natural explanations. ID is reliant upon forces acting outside of the natural world, forces that we cannot see, replicate, control or test, which have produced changes in this world. While we take no position on whether such forces exist, they are simply not testable by scientific means and therefore cannot qualify as part of the scientific process or as a scientific theory.

Judge Jones was extremely clear about the lack of any scientific approach in intelligent design, but he did not stop there. Because the entire problem with teaching intelligent design is one of separation of church and state, or one of maintaining the Establishment Clause of the U.S. Constitution, the judge had also to address this problem with respect to intelligent design and the Dover school board. What he concluded was amazing. First, he asked what, as it is not science, intelligent design actually is. His answer was as follows:

The facts of this case make it abundantly clear that the Board's ID Policy violates the Establishment Clause. In making this determination, we have addressed the seminal question of whether ID is science. We have concluded that it is not, and moreover that ID cannot uncouple itself from its creationist, and thus religious, antecedents. Both Defendants and many of the leading proponents of ID make a bedrock assumption which is utterly false. Their presupposition is that evolutionary theory is antithetical to a belief in the existence of a supreme being and to religion in general. Repeatedly in this trial, Plaintiffs' scientific experts testified that the theory of evolution represents good science, is overwhelmingly accepted by the scientific community, and that it in no way conflicts with, nor does it deny, the existence of a divine creator.

Intelligent design is thus nothing more or less than creationism in disguise, but Judge Jones didn't stop there. He also asked if the Dover school board was honest in its attempts to propose ID as a scientific approach to origins on this planet. If board members honestly thought that ID was scientific, but were misled or had misinterpreted ID as scientific, then the judge might have taken another tack. But he found that:

The citizens of the Dover area were poorly served by the members of the Board who voted for the ID Policy. It is ironic that several of these individuals, who so staunchly and proudly touted their religious convictions in public, would time and again lie to cover their tracks and disguise the real purpose behind the ID Policy.

Finally, in detailing all of this, Judge Jones ended his decision by filling a hole in the decision that is almost always raised in legal actions such as the Dover trial. Many judges are accused of being activists and not judges. This is because some judges can impose their own ideas on the public by their decisions, and, in essence, actively influence society rather than merely apply the law. Judge Jones makes it clear that he is not an activist:

> Those who disagree with our holding will likely mark it as the product of an activist judge. If so, they will have erred as this is manifestly not an activist Court. Rather, this case came to us as the result of the activism of an ill-informed faction on a school board, aided by a national public interest law firm eager to find a constitutional test case on ID, who in combination drove the Board to adopt an imprudent and ultimately unconstitutional policy. The breathtaking inanity of the Board's decision is evident when considered against the factual backdrop which has now been fully revealed through this trial. The students, parents, and teachers of the Dover Area School District deserved better than to be dragged into this legal maelstrom, with its resulting utter waste of monetary and personal resources.

In other words, thanks for wasting my time, your time, and your children's time, but it won't happen again on my watch. Judge Jones was, by the way, appointed to his first government job in Pennsylvania by Republican Gov. Tom Ridge and then to the federal Court by President George W. Bush. He has had a record of conservative decisions in both posts, so it's hard to argue that he had an ideological axe to grind here. But the story doesn't quite end even there. Four days after testimony in the trial ended, an eight-member slate of candidates opposing the Dover ID resolution was voted onto the Dover school board. Earlier members who had supported the resolution were resoundingly ousted from the board membership.

Even More Important

Evolution is much more than a theory that helps us answer questions about the origin of human beings and the ever-changing nature of life on Earth—topics that are typically the focus of debates on the subject. For while these debates are important in their own right, support for Darwin's theory is bolstered by demonstrating the impacts of evolution on our lives. Modern humans are, by and large, practical people, and the practical implications of evolutionary theory on our everyday lives also shed some light on why evolutionary theory is so important. Evolutionary principles, and the predictive value of studying evolution, influence our lives in at least three major ways—in human disease research, in bioterrorism preparedness, and in energy production. [Figure 8]

First, the Human Genome Project, the multibillion-dollar international effort to sequence, interpret, and exploit the human genetic code, illustrates the significance of evolutionary theory. The field of comparative genomics enables scientists to identify genes and their functions by comparing the DNA sequences of two or more species that share a common ancestor. This is why the sequencing of more than 1,000 nonhuman genomes—from mammals to bacteria to plants—has been such an integral part of genomics. This evolutionary information helps scientists and doctors develop innovative and effective diagnostics and treatments for diseases such as cancer, diabetes, and heart disease.

Evolutionary theory enhances our ability to understand pathogens that not only might occur naturally, but that also could be used in a bioterror attack. In the event of such an attack, the rapid identification of pathogen strains using a comparative evolutionary approach (similar to that used by the Human Genome Project) has the potential to be helpful in developing countermeasures and treatments.

Evolutionary principles are being used to decipher the genomes of microbes responsible for infectious diseases. Evolutionary biology has revealed the incredible abilities of microbes that cause everything from Severe Acute Respiratory Syndrome (SARS) to avian influenza to adapt to their environments and develop deadlier strains. An evolutionary perspective helps scientists and clinicians understand how both antibiotic resistance and changes in the genes of microbes can enhance a microbe's evolutionary success and thus threaten human health.

Finally, evolutionary biology has opened up opportunities to study Earth's microbes as a possible source of clean energy. It is no wonder that J. Craig Venter, who as president of Celera Genomics led its sequencing of the human genome, is now dedicating much of his research to using comparative evolutionary techniques to develop environmentally friendly microbial energy sources.

We are worried about the impact on our nation's or any nation's scientific, medical, and public health infrastructures were the Darwinian foundations of biology to be surrendered to an ideology of creationism. Ultimately, this is not a debate about science versus faith; faith and science do not need to be exclusive of one another. After all, faith coexists quite happily with other scientific fields like physics and chemistry. Instead of debate, scientists and supporters of evolution should educate, helping the public understand both the nature of science and the basic tenets of evolutionary theory as well as the important discoveries that are made possible by the application of evolution to biology.

FIGURE 8. Earth's microbial diversity and evolution. A Black Smoker, source of much extremophile microbial life.

If as a society we abandon the teaching of evolution and origins in our schools, we will not only leave our children behind, but we will devastate our scientific and public health infrastructures. This will halt progress and hope on humanity's scientific frontier—and that, no amount of faith would be able to restore.

We have digressed into this important area of the role of evolutionary thought in our everyday lives because it shows the importance of understanding origins. Origins are at the heart of understanding evolution and ultimately, we believe, at the heart of understanding the human condition. It is important to think back to all of the attempts by recent humans to grapple with the almost overwhelming subject of our origins, and to view current evolutionary thought and ideas about our human origins as yet another step in becoming more human. In the following chapters we will delve into the tools that help us take the scientific journey to an understanding of our origins.

CHAPTER TWO

PALEO-ANTHROPOLOGY

W E'VE ALREADY SEEN HOW PALEOANTHROPOLO- GY IN ITS MODERN GUISE GOT ITS START AROUD 1960, WHEN LOUIS AND MARY LEAKEY CO-OPTED A

variety of specialists to work with them as they excavated the fossil human-containing deposits of Olduvai Gorge, in Tanzania. This was actually quite a turnaround for Louis, who up until then had been the personification of the "lone paleontologist." But the Leakeys had obviously hit on the wave of the future.

The collaborative theme was soon taken up on a larger scale by the American anthropologist Clark Howell, when he organized a series of field expeditions to the remote wastes of the Omo Basin in southern Ethiopia in the late 1960s. A Howell expedition could easily consist of a couple dozen members, including several Ph.D.s and, as likely as not, a helicopter, in addition to a fleet of Land Rovers. Since then, paleoanthropology has in many ways become Big Science, like genomics, and it often seems that no expedition is complete without a host of specialists in tow and often, given the wild and remote places in which human fossils are often found, a gaggle of armed guards.

Diverse as paleoanthropology is, its hard core is still the human fossil record, and any description of the science has to start with fossils, which usually consist of petrified bones and teeth, the hardest tissues in the body. Still, it is not easy even for something as tough as a bone to become a fossil on a paleontologist's workbench. [Figure 9] First of all, an animal has to die in the right place, that is to say, somewhere that it will not only escape complete obliteration by scavenging beasts, but where it will be covered by protective sediments—mud from flooding rivers, say—before wind and weather complete the destruction process. Then, the remains have to escape annihilation underground by chemical action or earth movements. After that, they have to be exposed once more by erosion of the rocks above them, and be collected by paleontologists before the elements have a chance to destroy them—which is not a good bet, for rocks containing many kinds of fossils, especially fossils as rare as human ones, are hard to find.

Even where you have located promising rocks, paleontological fieldwork can be tough indeed, especially when carried out in rough and remote country. Frequently, it involves surveying vast areas of scorching desert on foot, searching the rubbly surface for the slightest hint of bone. Worse, it is often unrewarding. Unlike all of the technical advances that have taken place in genomics laboratories in the past two decades, paleontological fieldwork is still done the old, hard way, although technology has intruded to the extent that findspots can now be pinpointed by the Global Positioning System, and satellite images of Earth's surface are now beginning to be used to identify promising rock outcrops.

Once the fossils are back in the laboratory, you have to clean them of the matrix with which they may be covered, and prepare them for analysis. This is often not easy;

FIGURE 9. The life history of a fossil. After death, most vertebrates will be devoured by predators or scavengers (top left). What is left over will either weather away or become buried in accumulating sediments (top right). Under appropriate conditions such buried remains may become fossilized, their constituents replaced by minerals (bottom left). If erosion subsequently wears away the sediments above, the fossil will be exposed at the surface again (bottom right). It must then be collected before it is destroyed by the elements. Illustration by Diana Salles.

adhering matrix may be rock-hard and unwilling to come away from the bone, and the fossils themselves more often than not are fractured and crushed, their pieces needing to be restored to their original relationships with one another. This can be an exacting task, and sometimes risky for the fossils themselves.

Braincases pose a particular problem, because they may be filled with tough sediment that somehow has to be removed to reveal the internal anatomy of the skull; if it was not already cracked, the only way of doing this used to be by physically breaking

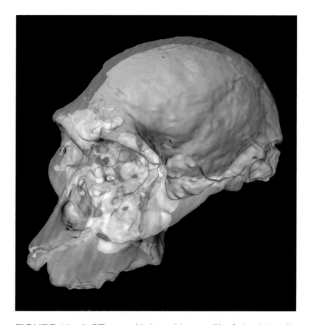

FIGURE 10. A CT scan ("virtual image") of the 2.5 million year-old *Australopithecus africanus* cranium Sts 5 from Sterkfontein, South Africa. The bone is shown in translucent gray, while the endocranial cavity (the space formally occupied by the brain) is shown in blue. Sascha Senck, Department of Anthropology, University of Vienna; courtesy of Gerhard Weber.

the braincase open, then laboriously removing the matrix from the interior using dental tools, vibrating needles, and other devices. However, one of the most exciting developments in recent years has been the development of a "virtual anthropology," using computerized tomography scanning, originally developed for clinical use. [Figure 10] CT images of fossil braincases and other bones are often made in hospitals, usually late at night when there are fewer patients to be imaged, because patients always get priority over even the most important fossils! The scanning procedure allows three-dimensional computer images of fossils to be generated and then manipulated at will. It is possible, for example, to "digitally subtract" the matrix without having to remove it physically.

In the case of a cranium, say, this will reveal the anatomy within that is not visible from the exterior, including sinuses inside the bone that cannot be reached by any physical means. It also allows accurate measurement of the skull capacity (a proxy for brain size), even where the braincase is still clogged with sediment. Further yet, the CT images can allow the different bits of a fractured skull (or whatever) to be virtually separated out and reassembled to provide an accurate view of what the specimen looked like when it was fresh, unbroken, and undistorted. To cap it all, the computer image can then be converted to an accurate and tangible model of what the original skull looked like, using "3-D printing" techniques originally used to produce models of broken or anomalous bony structures to help surgeons plan corrective operations.

Inferring Environments

Almost invariably, the ancient world that the fossils lived in was very different from the world in which they are discovered by paleoanthropologists; and there are whole subdisciplines of paleoanthropology devoted to understanding how those ancient worlds can be reconstructed from the meager indications that the paleontologist finds. This can very often be a tricky business, because the fossil animals that a paleontologist finds in a particular place are normally not a typical sample of those that lived there in the remote past. For although the "death assemblages" that a paleontologist deals with are derived from a once-living fauna, they typically consist simply of those species and bony elements that are most likely to be preserved after an often rough ride. For example, bones that eventually become fossils are often transported quite considerable distances before reaching their place of burial. Usually they are moved by water—swept along in floods, or moved downhill by heavy rains.

Sometimes, other agents of accumulation are involved. Thus, hyenas were apparently responsible for the preservation of a large number of important human fossils. Hyena dens are places to which adults bring back parts of animal carcasses to feed their offspring. Moreover, they are often places in which any accumulated bones are subsequently protected from erosion. Leopards, too, have been important hominid bone accumulators, as a result of their habit of stashing their prey in favorite trees. On rare occasions, these trees happen to be growing in humid cracks in the ground that lead down to underground cavities, within which falling body parts may be preserved. At one South African site, a partial skull of an early human was found bearing a pair of holes that perfectly match a leopard's dagger-like canine teeth! Even porcupines have been implicated in ancient human bone accumulations. The upshot of all this is that if you want to reconstruct an ancient human milieu, you have to be able also to reconstruct the history of the fossil bones after the deaths of their owners. This is the realm of a group of specialists known as taphonomists, who study exactly what can happen to an individual after death, and how that history inscribes itself on the remains.

Rarely is a human fossil found in isolation. Mostly human fossils are found as a rather small part of a larger fossil fauna that, to a greater or lesser degree, samples the animals that shared the environment with them in life. Different faunas are characteristic of different kinds of habitats, with the result that not only are paleontologists of other specialties needed to identify the species concerned, but paleoecologists are needed to help decipher what those animals are telling us about what the ancient environments were like. Geologists of various kinds may also contribute to paleoenvironmental reconstruction. From the general nature of the sedimentary rocks in which fossils are found, it is possible to say a lot about the general setting in which those rocks were laid down, and microscopic examination of ancient soils also yields valuable information about the environments in which they were formed.

Dating the Past

But the most important subspecialty of geology involved with paleoanthropological fieldwork is stratigraphy, the study of the sequences in which the sedimentary rocks that often contain fossils were laid down. [Figure 11] Younger rocks lie on top of older rocks, so the higher in a rock pile the fossils occur, the younger they are. But not so fast: although one of the basic rules of stratigraphy is that sedimentary rocks are laid down in horizontal layers, subsequent earth movements can play havoc with the original "layer-cake" arrangement, and, in some cases, strata can even become folded over so that the older fossils lie above, creating a nightmare of reconstruction for the stratigrapher. Another problem the stratigrapher faces is correlating rocks of the same age from one area to another. Sedimentary rocks are formed from particles that are eroded from older rocks and then washed or blown away, to settle eventually in the bottom of depressions known as sedimentary basins. Each sedimentary basin has its own unique history. Traditionally, fossil faunas were used to correlate between basins, on the principle that faunas are not only characteristic of particular places, but of particular times, as well.

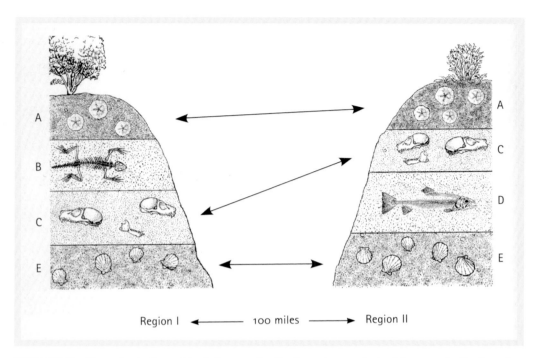

FIGURE 11. How biostratigraphic dating works. Sedimentary rocks build upward with time, but the record is seldom complete in any one place. This diagram shows how fossils can allow incomplete sequences from different areas to be integrated into a more comprehensive scheme. Illustration by Diana Salles.

Lately, a new approach has become available that can be used under certain circumstances and has proven extremely useful in unraveling the details of certain times and places in human evolution. This is "geochemical fingerprinting," which depends on the fact that the products of individual volcanic eruptions carry a unique chemical signal. Because volcanic ash can be blown over very wide areas before settling on Earth's surface and being incorporated into the accumulating rock record, it has been possible to tell that ashfalls found in places thousands of miles apart resulted from the same volcanic eruption and are therefore of the same age! Such marker beds provide a way of aligning rock sequences over enormous areas.

Faunal correlation permitted geologists to develop "relative" chronologies—rocks with faunas of this kind are older or younger than ones containing different faunas—and a reliable general sequence of periods of Earth history was worked out on this basis. [Figure 11] But until means became available for dating rocks in years, there was no way of accurately calibrating this succession. There was also vast uncertainty in correlating rocks separated by very long distances because, for instance, a tropical fauna looks very different from a temperate one of the same age. Enter the chronometricians, the geologists and geochemists who over the past half-century have developed methods of accurately dating certain kinds of rocks. In recent decades, a whole slew of chronometric dating methods, which yield dates in years, have become available. Most of these date the sediments (rocks or archaeological deposits) in which the fossils are found, but especially with recent advances in technology the most venerable of them, radiocarbon dating,

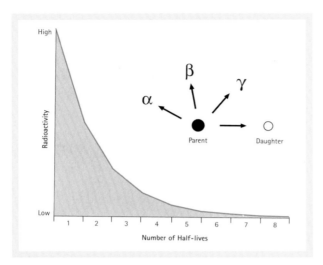

FIGURE 12. Radioactive decay. Radioactive isotopes (unstable forms of elements) "decay" into more stable "daughter" forms by emitting radiation or particles or both (upper right). The rate of such decay is expresssed as the "half-life," the time it takes for one-half of the atoms in a system to decay. The result is an exponential decay curve (lower left), which drops rapidly at first, then declines more slowly. Illustration by Diana Salles, after Tjeerd Van Andel, *New Views on an Old Planet: A History of Global Change*, 2nd edition (Cambridge University Press, 1994).

can sometimes be used directly on human fossils, in addition to other organic remains.

Unfortunately, radiocarbon dating can only be used on specimens that are less than about 40,000 years old. [Figure 12] Beyond this age, a variety of chronometric techniques is available, each of which has its own specific requirements. Nearly all chronometric methods depend in one way or another on radioactivity, the process by which "unstable" atoms "decay" to achieve a stable state. This happens at a steady rate that is conventionally quoted as the "half-life," i.e. the time it takes for half the atoms in a system to decay.

Radiocarbon dates are determined by measuring how much is left of the original unstable form; other approaches measure the accumulation of the "daughter" products of decay. The most widely used example of accumulation dating is the potassium/argon method (K/Ar, recently metamorphosed into a variant known as Ar/Ar dating), which exploits the decay of radioactive potassium into the noble gas argon. This technique dates mostly volcanic rocks, which are particularly good stratigraphic indicators because they are typically laid down over short periods of time. Of course, volcanic rocks are not found everywhere, but where they are, they are a godsend to a stratigrapher or paleoanthropologist, especially where they bracket the sediments in which a fossil is found. If fossil-rich sediments are interlayered with volcanic rocks, as is often the case in places like eastern Africa, the fossils will be slightly older than the dated rocks above them, and younger than those below. Since the half-life of radioactive potassium is very long, the K/Ar method can be used to date very old rocks indeed, but with recent refinements it can now also be used on rocks as young as a couple of hundred thousand years or even less.

A third basic approach is furnished by "trapped-charge" dating. This measures the numbers of free electrons that are trapped in defects in the crystal structures of minerals that are associated with fossils, and the number of electrons trapped is again a function of time. Materials in which this can be done range from flint to redeposited lime to dental enamel, all of which may be found in association with fossils, and which, in the case of the last, involve fossils themselves. Examples of this approach include thermoluminescence (TL) dating, for which flint tools and fragments burned in campfires are favorites, because heating empties the electron traps and resets the "clock" to zero, and electron spin resonance (ESR) dating, which can be used on fossil teeth as well as on the flowstones that sometimes cap fossiliferous deposits in limestone caves. Both of these approaches have proven very valuable in dating archaeological deposits and materials; TL works well back to 200,000 years or so, and ESR very much farther.

The emergence of chronometric methods has revolutionized our understanding of the timing of events in human evolution. When, in 1961, Jack Evernden and Garniss Curtis of the University of California at Berkeley used K/Ar to date Louis Leakey's first human fossil finds from Olduvai Gorge, Leakey himself was guessing that his fossils were maybe 600,000 years old. When the K/Ar date came in at three times that age, not only Leakey himself but the entire profession was flabbergasted!

Forty-five years down the line, we can be confident that we have, by now, a pretty good handle on the general timing of major events in human evolution, although some individual dates are pinned down remarkably tightly while others are bracketed by wide ranges of uncertainty. And since new instrumentation and new chronometric methods are coming on-stream all the time, we can hope that eventually we will be able to derive accurate dates on all of the fossils we find, rather than merely a large portion of them. But even then, the exact ages of some fossils found long ago may well always remain a mystery.

Identities and Relationships

With luck, you know how old your fossils are and, at least in general terms, the kinds of environments they lived in. Now you need to know, if you don't already, what species those fossils belonged to and how they are related to other species in their groups. For, although many paleoanthropologists scorn it as "an argument about names," knowing what to call your fossils is not only the most fundamental piece of knowledge you need to have, but is also one of the toughest things to figure out.

All living organisms, and thus all fossils, belong or belonged to a species. Unfortunately, exactly what species are is one of the most hotly argued topics in all of biology. In a general way, it is agreed that species are reproductively limited groups: As the basic "packages" that exist in nature, they are the largest populations within which all members can at least potentially interbreed. But after that, all hell breaks loose. From a paleontologist's point of view, there's no way you can tell reproductive behavior from looking at a fossil, while in the living world, some groups that look like perfectly good species may interbreed and even produce apparently fertile offspring. A striking recent example of this phenomenon is provided by polar bears and grizzly bears—two species that have been recognized as such for a long time. Every once in a while, a polar bear shows up with brown patches of fur on a normal snow-white pelage. When these bears are examined more closely, they also show the long claws and humped back of a grizzly bear. Because the polar bear and grizzly bear have different gene sequences, by examining their genetics one can clearly determine that these funny-looking polar bears are, in fact, the result of hybridization events between polar bears and grizzly bears.

So, while there can be no doubt that Nature is packaged in some meaningful way, this packaging is evidently very untidy; and as a result, at our last count scientists were using well over 20 different theoretical definitions of what species are. If there is that much division in theory, imagine what differences there are in practice! Still, even allowing for the undoubted fact that all species are variable (in the sense that their members differ from each other in a host of features), it is clear that the recent tendency among paleoanthropologists has been to minimize the number of species in the human fossil record.

Before we explain what we mean by this, some background: Closely related species are grouped into units called genera. This gives each species a two-part name: The first part is the name of the genus; the second is that of the species within it. This is why we identify ourselves as *Homo sapiens*, the species *sapiens* of the genus *Homo*.

In the years before World War II, practically every new human fossil that came along was given its own species name. Not unusually, it was given its own genus name as well, rather as each of us has given and family names. For example, in 1928, a skull found in 1921 at the site of Kabwe, in what is now the African country of Zambia, was given the name of *Cyphanthropus rhodesiensis*. When the Modern Synthesis came along in the mid-20th century, its proponents pointed out the critical fact that all animals belong to variable populations, and that fossils did not deserve new species names simply because they were not identical to something else already known. As a result, most authorities today would classify the Kabwe skull not only in our own genus *Homo*, but in the species *Homo heidelbergensis*, meaning that it doesn't even differ at the species level from fossils already described from elsewhere. Now, the notion that all populations are variable and that this has to be taken into account when classifying new fossils was an entirely laudable point to make. But as with all good ideas, this one was taken to a crazy extreme. At one point, under the influence of the Modern Synthesis, paleoanthropologists were cramming the whole human fossil record into a mere three species (*Australopithecus africanus*, *Homo erectus*, and *Homo sapiens*). As more and more fossil humans became known, this minimalist scheme rapidly began to bulge at the seams, and the number of species recognized had to multiply. But in our view, the pendulum has not yet swung far enough. We are still recognizing too few human species (or even, probably, genera) to properly reflect the amazing anatomical variety that we see in the human fossil record.

The big practical problem in reaching agreement on how many species are represented in the human fossil record lies in the fact that, as we'll see in more detail later, two major elements are involved in the evolutionary process. On the one hand lies structural change, and on the other is speciation, the origin of new reproductive communities. These events, as it turns out, don't necessarily occur at the same time. If you look around the living world today, you can find widespread species that have developed an enormous amount of variation within and among local populations that are still more than happy to interbreed if given the chance. But at the same time, you can find populations that are reproductively distinct that you can barely, if at all, tell apart by eye. When you're dealing with fossils, all you have to go on is what they look like; and even if you are making apparently commonsensical comparisons based on the amount of physical difference you tend to find among similar species in the living world, there is still plenty of room for argument. So argument there is, aplenty.

The family tree of humans and pre-humans contains just those species that most (though not all) paleoanthropologists would recognize as valid today. To us, there could well be more; but the point here is that until you have general agreement on how many evolutionary units you are dealing with, it is hard to come up with a coherent story, just as it's hard to make sense of what's going on in a play if you don't know who all the actors are. This is a major reason why paleoanthropology is such a famously argumentative science. Still, with each argument we inch a little closer to a more accurate description of the past worlds we see so fuzzily reflected in the available record.

Analyzing Evolutionary Histories

As we've just noted, by around 1950 the Modern Synthesis was preaching the notion that there were a mere three species in all of known human and pre-human history. What's more, these species were arranged into a single, gradually modifying lineage. But early in the 1970s a realization emerged that the overwhelming signal in the evolutionary record was one of stability and discontinuity, rather than of steady change. Species originate and then tend not to change much, at least directionally, until, after a variable tenure, they finally disappear—often to be replaced by a close relative.

The story of life on Earth thus appeared not as a gradual unfolding, but as an episodic affair in which the origin of new species, and competition among them, has played a key role. This has turned out to be as true in the case of fossil humans as elsewhere, and the notion of human evolution as a single-minded slog from primitiveness to perfection has been replaced by a much more interesting story, in which successive radiations of hominid species have competed for space on the ecological stage.

This has had major implications for the ways in which evolutionary histories have been reconstructed. [Figure 13] For if evolution consisted purely of gradual generation-by-generation change under the beneficent hand of natural selection, then figur-

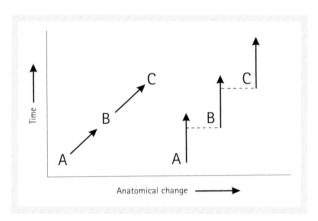

FIGURE 13. Two views of how evolution occurs. Represented on the left is "phyletic gradualism," whereby species gradually transform over time into other species. In contrast, the notion of "punctuated equilibria" (right) sees change as episodic, species being essentially stable entities which give rise to new species in relatively short-term events. Illustration by Diana Salles.

ing out evolutionary histories was largely a matter of discovery. After all, if fossils were more or less like links in a continuous chain, then to know more about the evolutionary history of any group, all you needed to do was to go out and find more of them, and their age would tell you pretty much where they fitted in. But if species as wholes in fact play a crucial role in the evolutionary process, and if the evolutionary history of any successful group tends to consist not simply of successions of slowly changing species but of changing diversities of coexisting species, then understanding those evolutionary histories became a matter of analyzing the relationships among those species, rather than one of simply discovering the links in a chain. And as it happened, more explicitly grounded techniques for doing this were becoming available as the 1970s began. In particular, a method known as cladistics was adopted. It involved recognizing relationships between pairs of species on the basis of innovations inherited from their most recent common ancestor. The pairs could then be grouped with others, using the same criterion, until you had a large branching diagram that expressed the relationships among all the members of a large group, ultimately radiating out to include all living things.

But all that a cladogram does is to express the closeness of relationships among species, based on recency of common ancestry. It says nothing about ancestry and descent, which are, in fact, more difficult, if not impossible, to demonstrate. If you add these in, you are moving away from what Karl Popper would call testability, but you are getting closer to a more interesting formulation – the more familiar "evolutionary tree," in which you plot both the relationships and the geological age of fossil species. The tree can then form the basis for a narrative account known as a "scenario," in which you throw in everything you know about adaptation, ecology, behavior, and so forth. Popper would say that this is the least testable proposition of all, but it is at the same time the most interesting – and it is just fine to dream up scenarios, provided you are explicit about the evolutionary tree and, most importantly, about the testable cladogram on which it is based. Forty years ago, paleoanthropologists just dived in at the deep end with the scenario; now many of them are acutely aware of the need to get there the hard way, by doing the prior analyses. This, of course, brings us back to the genomic trees we introduced earlier, which are conceptually a slightly different creature, because they are based on the comparison of (almost exclusively) living forms.

Fossils as Living Creatures

So now we know who our fossils are, where they are from, how old they are, what kind of environment they lived in, and what other animals they shared that environment with. An obvious next question to ask is exactly how they made their living.

When we're dealing with ancient humans, there is a number of approaches to this. One is to study how our precursors functioned as mechanical contraptions. We use the word "contraption" advisedly, because although, for example, we like to think of ourselves as well adapted to the upright walking that comes to us so naturally, a lot of compromises were in fact involved in converting an ancestral quadruped into the striding bipeds we are today. Our high frequencies of slipped discs, fallen arches, and wrenched knees are only a few of what have been called the "scars" of human evolution. Nonetheless, like all other successful organisms, we are pretty good at what we do, and our adaptations for our unusual way of moving around are evident in our skeletons. This means that, to a large extent, we can read back from the structure of ancient bones to how their original possessors had used them in life. Functional anatomists are adept at doing this, and it turns out that by studying such things as limb proportions and the shapes of individual bones and bony complexes, it is possible to get a pretty good idea about the way fossil forms moved around when they were parts of vibrant living ecosystems.

A very important part of making a living is what you eat, and this is something that can be read very broadly from an organism's teeth. [Figure 14] There is no mistaking a carnivore for a grass-eater, for example. More subtle variations in diet are less easy to read from the form of the teeth, but close scrutiny can still tell you a lot; this is particularly true if you add the fact that teeth wear as they are used, and that at the microscopic level the traces of wear on your teeth may be very different, depending on the

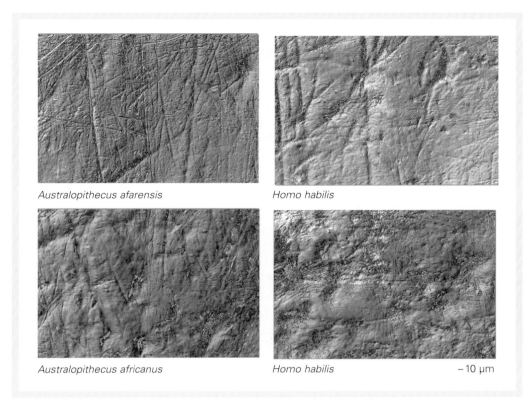

Australopithecus afarensis

Homo habilis

Australopithecus africanus

Homo habilis — 10 µm

FIGURE 14. Confocal micrographs at high magnification showing areas on the crushing/grinding facets of the molar teeth in four different kinds of hominids. The gouging and pitting seen on the *Paranthropus* tooth indicates a diet of tough and possible gritty objects such as seeds or tubers; the striations on the other teeth suggest a predominance of softer foods such as fruit and possibly leaves. The scale is 10 microns. Courtesy of and © Peter Unger.

substances you are masticating. If you look at the wear-surfaces of teeth at high magnifications using a scanning electron microscope, for example, you find that the striations and pitting seen on, say, the chewing teeth of an animal that eats roots and tubers, are very different from anything you see on the teeth of fruit-eaters. Knowing what you see on the teeth of living forms with known diets helps you to read back to what ancient animals were eating.

Moreover sometimes, particularly among ancient humans, tooth-wear can tell you about more than just diet. Occasionally teeth are used as tools, to hold or process things that are not eaten; and that shows up in tooth-wear too. Neanderthals, for example, typically show extremely worn front teeth, worn in a very characteristic way that suggests they may have been used to process hides, or to hold them as they were being held taut with one hand and scraped with a tool held in the other.

Conversely, teeth aren't the only thing that can tell you about diet. Isotopic chemists have recently begun using the ratios of certain stable isotopes (alternative forms of elements, particularly of carbon and nitrogen) that may be preserved in fossil bone, to help clarify the diets of ancient humans. They have come up with some surprising results, including the suggestions that certain very early humans may have eaten more meat than had been supposed, and that some Neanderthals, at least, were very heavily carnivorous, possibly even specialists on such fearsomely large herbivores as woolly rhinoceroses and mammoths. The abundance of some trace elements in bone may also be useful as dietary indicators; at present, both stable-isotope and trace element studies are in their infancy, but chances are that we'll see rapid developments in the near future.

As far as reconstructing ancient behaviors is concerned, though, paleoanthropology has one huge advantage over studies of other extinct creatures. Humans alone have left us a direct material record of their ancient behaviors, rather than merely their bones. This is the archaeological record and it starts some 2.5 million years ago, when human precursors began to make stone tools. Because they are readily recognizable and are more or less indestructible, stone tools are abundantly known. What's more, a series of technological innovations over the time since stone tools were first made gives us some insight into the cognitive advances made by successive kinds of early humans.

The period of human prehistory covered by this book lies more or less entirely within the Paleolithic period, or the Old Stone Age (roughly 2.5 million to 10,000 years ago). This is a long expanse of time, during which ancient humans were gatherers and scavengers, then hunter-gatherers, always pretty consistently on the move.

Not until close to the end of the Paleolithic did human precursors begin to build shelters, at least of the kind that might be expected to be preserved, and so most archaeological sites of the period are pretty modest. The earliest sites are, indeed, no more than spots on the landscape where early humans butchered animal carcasses that they had more probably encountered than hunted. With time, Old Stone Age archaeological sites become a bit more complex; and early humans began to return repeatedly to favored camping spots, often accumulating large thicknesses of archaeological deposits over the millennia. Such deposits consist basically of what early people threw away, or simply left behind. Not for nothing has archaeology been called "the study of ancient garbage."

Still, there's a lot you can learn from ancient garbage (and from modern garbage too, as the "garbagologists" who trawl through the trash of celebrities can attest: People's garbage does not always confirm what their PR agents say about their lifestyles). Moreover, you can learn not just from the garbage itself, but from where it is found. The places where archaeological sites are situated, and the ways in which they were distributed across ancient landscapes, tell archaeologists an awful lot about the lifestyles of the people that made them. Even the tools themselves are not useful simply as witnesses to the development of human technology, but have broader implications about the lives of the people who made them.

Sadly, stone tools are only part of the technological story, and they give us only a very indirect glimpse of the full toolkit of the people who made them, still less of their intelligence or of the way they perceived the world, or of the social aspects of their lives. Nonetheless, its durability makes the stone tool record a very complete record of one limited aspect of human activity. For this reason, even where archaeologists are essentially limited to the contemplation and characterization of ancient stone tool "industries," they often broaden these designations to the cultures and societies that made them.

For example, the first period of modern human occupation of Europe is known as the "Aurignacian." Technically, the Aurignacian is defined by a characteristic assemblage of stone tools (with the addition, in this case, of a slender split-based spearhead made of bone). Strictly speaking, then, it is the technological assemblage that is Aurignacian, not the people who made it. But archaeologists liberally speak of "Aurignacian people," on the perfectly reasonable assumption that making the diagnostic tools of the Aurignacian industry was something learned, and that this learned knowledge was passed on within societies that were united by many more cultural elements than simply these particular stone working techniques. In this sense, it is entirely justifiable to talk of an "Aurignacian culture."

Like paleoanthropology itself, as Paleolithic archaeology has evolved it has co-opted a range of specialists into its activities. The lead archaeologist on a site will typically be a generalist, organizing and coordinating its excavation, which will be done by individual excavators, frequently students, each of whom has responsibility for a particular part of a site. The excavated area will normally be defined by a grid against which the position of everything found is mapped in three dimensions. Once the material is discovered, the specialists come into play. Archaeometrists concern themselves with dating archaeological deposits, using many of the dating methods we described earlier, as well as technological comparisons and their own version of stratigraphy – for archaeological deposits accumulate one on top of another, just as do geological sediments. Other scientists also look at the biochemistry of bones and perform physical and chemical analyses of artifacts of all kinds. Archaeozoologists study the animal bones found at each site, identifying the remains to discover the diet, subsistence patterns, and butchery practices of the people who lived there. Archaeobotanists study the much rarer plant remains found at sites, including seeds, fibers, and even hardened mud impressions where such things are found. Palynologists study the plant pollen found in archaeological deposits, which can reveal a great deal about the surrounding environment.

Even those who limit themselves to the study of stone tools come at their data from a variety of perspectives. Some of them are concerned with the "typology" of stone tools, categorizing the types of tools that were made in particular times and places; others with the techniques by which the tools were made, and by the "reduction sequences" by which a piece of fresh stone is converted into a brand-new tool and then resharpened until its shape is completely transformed; others by the ways in which the tools were actually used, and how those particular uses inscribed themselves on the tools that have come down to us. In this last connection, experimental archaeologists indulge in extreme behaviors, such as butchering entire elephants using tiny sharp stone flakes, just to see what is possible. New approaches are being discovered all the time, and indeed, the variety of archaeological preoccupations is as inexhaustible as human behavior itself.

What Does It Mean to Be Human?

Up to now we have been throwing around the term "human" pretty loosely, which is perhaps excusable because people have been describing themselves as human since long before anyone had the slightest idea that we had relatives out there as close as the apes, let alone much closer relatives that are now extinct. But it does beg the question of what, exactly, we mean by this vernacular term. If we could avoid this problem by retreating to technical jargon, that would be great. But it turns out that taking refuge in scientific terminology doesn't help much. One way of addressing what it means to be human, which we will explore in detail in Chapters 8 and 9, is to look at those things that make us unique relative to all other organisms. For now, though, we will use a second approach, which is to detail how humans are related to other organisms, living and extinct, on this planet.

Order:	Primates (Lemurs and lorises, tarsiers, monkeys, apes, humans)
Suborder:	Haplorhini (Tarsiers, monkeys, apes, humans)
Hyporder:	Anthropoidea (Old and New World monkeys, apes, humans)
Infraorder:	Catarrhini (Old World Monkeys, apes, humans)
Superfamily:	Hominoidea (great and lesser apes, humans)
Family:	Hominidae (humans and their extinct relatives)
Genus:	*Homo*
Species:	*Homo sapiens*

FIGURE 15. Classification of the human species. The rules of zoological classification produce an inclusive, rather than an exclusive, hierarchy, so that a taxon (group) belongs to all of the larger categories that lie above it. Thus our species *Homo sapiens* belongs to both the Infraorder Catarrhini and to the Order Primates.

Traditionally, modern human beings and their extinct relatives have been classified in the zoological family Hominidae, a division of the order Primates that also contains the apes, monkeys, lemurs, and bush babies. [Figure 15] All of the apes were classified together in the family Pongidae. But molecular studies back in the late 1960s began to suggest that, in fact, modern humans and their fossil relatives might actually be more closely related to one of the apes (most people's favorite candidate is the chimpanzee) than they are to the other apes. That, of course, placed the simple two-family dichotomy in doubt. Nowadays, the family Hominidae is often taken to include the chimpanzees, as well as humans and their extinct relatives.

In this book we prefer to sidestep the issue by adopting a "bottom-up" classification. There are now far more known extinct human relatives than there are species of living or fossil apes – so many species, indeed, that by any criterion this diverse group deserves family status. That is why we are sticking with the traditional use of Hominidae, and its derived adjective "hominid," to include just *Homo sapiens* and those extinct forms that are more closely related to it by recency of common ancestry than to any ape – though you will see plenty of alternative classifications elsewhere. We will look in detail at the place of Hominidae in the larger tree of life in Chapter 5.

Still, even with this under our belt we are left with ambiguity in the meaning of "human." And, to be quite frank, we ourselves are pretty sloppy about this, using the term differently in different contexts. When we speak of "human evolution," we and most others are referring to the evolution of everything within the family Hominidae, as used here. But from now on, when we use "human" as a descriptor, we will limit it to members simply of our genus *Homo*, only one of several hominid genera. Furthermore, the extinct species of *Homo* are not what we could properly describe as "fully human" in any functional sense: Our living species appears to be unique in the way it interacts with the world around it, and in a cognitive sense, only *Homo sapiens* can be characterized in this way. But that's not all. The very first *Homo sapiens* were "anatomically modern humans," because they looked just like us. But (to give away the plot early) they were not "behaviorally modern humans." This status, it seems, was only attained some time later. So while our choice of terms represents a consistent usage, reading the literature makes it evident that it's not an obligatory one. Finally, as to when "hominids" became "people," well, you just pays your money and you takes your choice.

CHAPTER THREE

WHAT'S IN A GENOME?

ONE MIGHT THINK THAT SEQUENCING THE HUMAN GENOME IS OF LITTLE OR NO CONSEQUENCE TO UNDERSTANDING OUR ORIGINS. IT'S ALL ABOUT THE

bones, right? But our genomes carry many footprints of our evolutionary past that our bones simply do not show. In addition, information from our genomes can complement what we know about bones.

Our genomes harbor two kinds of information. The first is a record of how we are related to other organisms, and especially to other human beings. In this case, because our genetic material is inherited from parent to offspring many of the changes in our genomes are a direct historical record of how our ancestors split into new populations and migrated into new areas of Earth (Chapters 5 and 7). The second is a record of the kinds of changes that make us unique or different from other organisms (Chapter 8). For instance, our genomes harbor genetic changes that have happened since our divergence from other great apes and are involved in the structure and function of our brains. Much of the information from our genomes is unique and tells us things that no other kind of information can. Our genomes, therefore, are an important part of our story here.

But wait! We haven't yet defined what a genome is. So the first important question we have to ask in this chapter is "What is a genome?" To answer this question, we need to look deep into our cells, because that's where our genomes reside.

A Fantastic Voyage

Before we get to all of this genomics stuff, let's take a little journey into a cell, to see what a genome is and why it is so important. This journey is an even more fantastic voyage than the one Raquel Welch (big surprise that she is the only actor we remember in the movie) took in the 1966 movie *Fantastic Voyage*. In that movie, the plot involved shrinking humans so they could travel through the blood stream of a colleague to cure him. In our voyage, we will progressively shrink ourselves to travel into the center of a cell to look at how molecular biology works and how genes, chromosomes and genomes are related to producing gene products—RNA (ribonucleic acid) and proteins. To make this voyage, we will focus on a single gene that is important in the ability of humans to perceive red and green colors. Don't worry if some of the things we see along this journey are strange and or foreign. We will explain much of what you see on this voyage later in the chapter and others. For now, sit back and enjoy the ride.

Our voyage starts outside the eye. As you get closer and closer to the eye, you begin to see different tissues. As you get closer to the tissues, you start to see cells. These

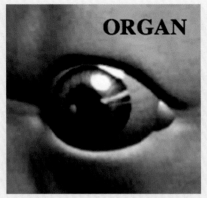

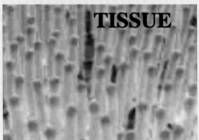

FIGURE 16. Top: External drawing of a human eye. Middle: Magnification of the surface of the eye showing the "field" of rods (longer cells) and cones (shorter cells). Bottom: Drawing of an even more magnified cone cell.

are strange-looking cells, because they look a lot like a shag rug up close. When we get a little closer to the shaggy-looking things, we see there are actually two kinds. [Figure 16] These are cells called rods and cones, so named because they look like little rods and cones. As you get closer to a cone cell, you start to make out the various parts of the cell. The most spectacular thing you can see about a cell is that it looks as if there is another cell or body within it. This large cell-like thing is called a nucleus. [Figure 17]

We are next going to go directly into the nucleus to see what is there. When we zoom into the nucleus, we start to see big stringy things that look like bundles of wire. These bundles are called chromosomes. [Figure 18] The cone cell of a human that we have zoomed in on has 46 of these chromosomes. We pick one chromosome to zoom in on even farther, and we start to see it unravel like a ball of string. [Figure 19] We realize that the chromosome is actually a long, long string with a definite molecular makeup. We start to zoom even closer, and we notice even smaller stretches on the chromosome, called genes. [Figure 20] We zoom on by the genes, and as we begin to magnify more, we start to see that each gene is actually made of two strings wound around each other like a spiral staircase. This is DNA. [Figure 21] In real life, we can't see the two strings when we are looking at chromosomes because they are so small, but on this journey, the magnification and our imagination allow us to see them.

At this point in our journey things get very curious indeed, because much of what we "see" is so small that we have to think of the things we "see" as abstract models. Sure, we can clearly see the arrangement of the atoms (mostly carbon, hydrogen, oxygen, and nitrogen) in DNA, but just seeing the atoms doesn't help us understand how DNA works. Instead we need to start to think about the shapes the atoms take along the DNA strands. Once we do this, we recognize that there are four major kinds of arrangements of the atoms on the DNA strands. These four different arrangements of atoms are called bases or nucleotides. Each base can be considered a single bit of information, much as a byte is considered the unit of information in computers. For now, we will call these four bases by their DNA names—G, A, T, and C—but keep in mind that these molecules are the most basic building blocks of our genome. Now we notice that the double-stranded DNA of

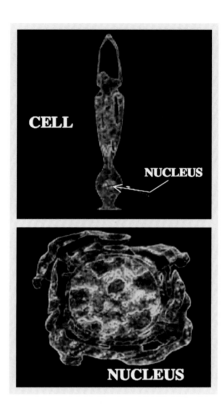

FIGURE 17. Top: Drawing of a cone cell showing the nucleus. Bottom: Drawing of the nucleus.

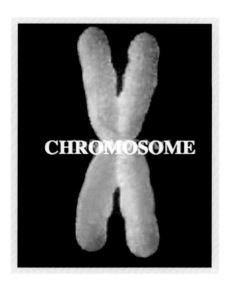

FIGURE 18. Magnified drawing of a chromosome.

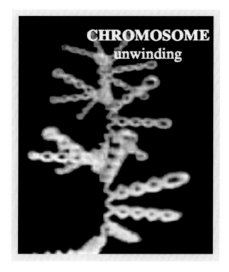

FIGURE 19. Magnified drawing of a chromosome tip showing the double helix in the chromosome unwinding.

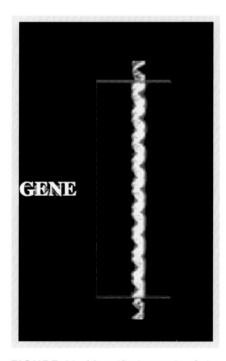

FIGURE 20. Magnified stretch of the double helix of the chromosome shown in previous figures, showing the location of a gene.

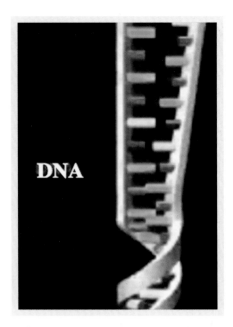

FIGURE 21. Highest magnification drawing of the double helix in the gene in the figure 20. While microscopes can "see" all stages prior to this one, current magnification processes cannot actually see the structure of the double helix at this level. Biochemical methods are used to visualize the double helix at this magnification. In this drawing, green is adenosine (A), red is thymine (T), yellow is guanine (G) and blue is cytosine (C).

FIGURE 22. Three dimensional model of a protein. The ribbons and threads represent stretches of amino acids. The curly ribbons are stretches of amino acids that make helices.

the gene we focused on—the one that helps us perceive red and green colors—has unraveled. If we look at it with our "molecule eyes," we see atoms, but if we look at it with our abstract "DNA eyes," we see strings of G's, A's, T's, and C's.

Once a gene unravels, another string of molecules is copied from the gene. This string looks a lot like one of the strands of DNA with the smaller molecules for G's, A's, and C's in it, but instead of T's in the string, there are U's. This string of nucleic acid is called RNA. The RNA then moves outside of the nucleus to a region in the cell called the cytoplasm, where machine-like objects called ribosomes come into contact with the RNA. The RNA string is ratcheted through the ribosome DNA base by DNA base, and through biochemical reactions a protein is synthesized from the instructions on the RNA. [Figure 22] Next, the protein comes off the ribosome and folds into a three-dimensional structure. The protein then starts to interact with light hitting the cone cell, and a signal is sent from the cone cell to the brain telling the brain what kind of light is hitting the cell.

Our journey is over. We have only looked at a single gene in a single kind of cell in a single kind of tissue. Imagine thousands and thousands of journeys just like this one with different genetic destinations, and you will have a good idea of our genome. The trick to modern genomics is to make all of these journeys to genes faster and easier.

Complex Journeys

Some of the journeys to view other traits become very complex. Color vision is a relatively simple genetic journey that we can make into our genomes. But what about journeys to the genes that control how our bodies develop, how tall we get, how we learn language, how we combat disease, or how we learn to play music? These other journeys are much more complex and involve visiting many, many genes at once. We are confronted with the problem shown in figure 23:

[Figure 23] shows a set of instructions, on the left, in the guise of DNA sequences giving rise to a complex behavior such as musical ability, on the right. We

```
ATGACTCAGGAGGAGGCTGGGCGGCTGCCCCAAGTACTGGCCCGGGT
CGGAACCTCCCACGGTATCACCGACCTGGCCTGCAAGCTCCGCTTCT
ATGACGACTGGGCTCCGGAGTATGACCAGGATGTGGCTGCTTTGAAG
TACCGAGCCCCACGCCTTGCTGTGGATTGTCTCAGTCGAGCCTTTCG
GGGCTCACCCCATGATGCCCTGATCCTCGATGTGGCCTGTGGCACTG
GCCTGGTAGCTGTGGAGCTGCAGGCTCGGGGCTTCCTCCAGGTGCAG
GGCGTGGATGGAAGCCCAGAAATGCTGAAGCAGGCGCGGGCACGTGG
CCTGTACCACCACCTTAGCCTCTGTACCCTGGGCCAGGAGCCACTGC
CCGACCCTGAAGGGACCTTTGACGCGGTGATCATTGTGGGTGCCCTC
AGTGAGGGACAGGTGCCCTGCAGTGCCATACCTGAGCTCTTAAGAGT
CACCAAGCCAGGTGGACTTGTTTGTCTGACTACCAGGACCAACCCAT
CCAACCTTCCATACAAGGAGACGCTGGAAGCGACCTTAGACTCCCTG
GAGCGGGCTGGAGTGTGGGAATGCCTGGTGACCCAGCCTGTGGACCA
CTGGGAGCTAGCAACATCAGAACAAGAGACAGGGCTAGGCACCTGTG
CCAATGATGGCTTCATCTCTGGCATCATCTACCTTTACCGGAAGCAG
GAGACAGTATAG
```

k

FIGURE 23. (left) Code for human genome. (right) Simone Dinnerstein performing Bach's *Goldberg Variations* at Carnegie Hall's Weill Recital Hall, November 26, 2005. Photograph by G. Paul Burnett/*The New York Times*/Redux

know from the past century of genetic research on traits like behaviors that a lot more is involved than just DNA sequences, and that genomes aren't everything. Genes interact to produce the outward attributes (or phenotypes) we see in organisms. Information about our genomes gives us a framework for understanding the intricacies involved in those "complex" attributes of anatomy and behavior that involve many genes and many gene interactions. We are finding that many of our most interesting characteristics are controlled not by a single gene but sometimes by hundreds of genes. How these genes interact with each other to produce our physical and behavioral selves is an important aspect of genomics.

And what about the nongenetic influences, or what scientists call the environmental component? Are there environmental aspects to the ways our genomes regulate the final structure of anatomical features and behaviors? Yes, of course. Understanding the complex interaction of the many genes that are interacting with the environment is no simple matter. While we can hope to understand our humanness from a genomic perspective, there are many caveats to an unwavering view of its being an all-genetic process. Although it is tempting to take the next step—to see ourselves entirely as a product of our genomes—we must also recognize the complexities involved in how genes work and integrate with our environments.

When the first draft (as this book went through many drafts, so too, the genome was completed in drafts) of the human genome was first announced, John Sulston, a British scientist deeply involved in the sequencing of the human genome, exclaimed, "We are to the point in human history where, for the first time, we are going to hold in our hands the set of instructions to make a human being."

The instructions that we can now hold in our hands on an iPod or on a CD-ROM are involved in producing the outward appearance and behavior of an organism or its phenotype. [Figure 24] We firmly believe that there is no phenotype independent of the influence of genes AND environment. There is a wide range of possible effects that are involved in the interaction of genes and the influences of the environment. But one thing is certain: There is no environmental effect without genes, and there is no genetic effect without environment. The two are intertwined and intermixed, and neither is able to exist without the other.

What Is It?

A genome is the collection of all the DNA an organism gets from its maternal and paternal parents. This collection of DNA comes from sperm of the father and egg of the mother and includes 25,000 genes or so, mixed in with lots of DNA that apparently doesn't do much, sometimes called "junk DNA."

So what, then, is a gene? The importance of our genes (and the genes of millions of other species) has been known for more than a century. Genes control aspects of development, metabolism, behavior, and other important characteristics of organisms. We know this from more than 100 years of research, not only on humans but also on several other creatures, such as the lowly worm or the irritating fruit fly. While not the most interesting living things in the world (these creatures are actually quite boring, except to the scientists who study them), these were chosen as study organisms in order to minimize the efforts of scientists in producing the most information about how genes work in controlling things. These boring little guys produce the bigger bang for the buck in the study of genetics.

To the pioneers of genetics like Gregor Mendel, genes were somewhat abstract entities. In fact, Mendel didn't even call them genes. The word "gene" wasn't used until about 40 years after Mendel died. To the workers who expanded the science of genetics during the first decades of the 20th

FIGURE 24. Digital media that could hold genomics data or be used for access to genomic data. If each base of the sequence of a genome occupies one byte then the CD-Rom disc could hold 70,000,000 base pairs of sequence, or about seven large bacterial genomes. As we were writing this book, Francis Collins appeared on the Colbert Report on television and gave the host of the show a DVD disc containing all of the bases of the human genome. The nano iPod with five gigabytes could hold 5,000,000,000 bases or a little under two human genomes' worth of information. The NYC Metro Card could be used as a key card to access a larger database holding genomic information from any number of genomes.

century, genes represented elements that controlled outward appearances of organisms, and were elements that had followed certain rules of behavior and inheritance. As we will shortly see, genes became more concrete with the discovery that they are made of DNA.

Because genes are made of DNA, and a genome is made of genes, your genome is made of DNA. We will see shortly that this genetic information can be examined in many different ways. For now, we will simply say that genes are made up of DNA, which is in turn made of long strings of four kinds of small molecules called bases. Each base in the DNA strings can be said to be a "bit" of information. As we have seen, any DNA string contains four kinds of these small molecules or "bits" of information, called adenine, guanine, cytosine, and thymine, or A, G, C, and T, for short.

These bases are small and cannot be seen with a microscope. Special biochemical methods are used to visualize DNA and the order of the G's, A's, T's, and C's on DNA strings. They have to be. The entire human genome has 3 billion of these bases or bits of information in it.

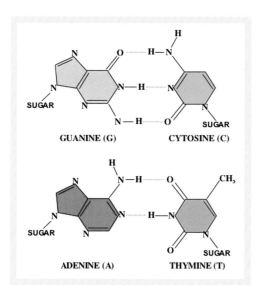

FIGURE 25. The four bases of DNA. Blue is guanine (G), yellow is cytosine (C), red is adenine (A) and green is thymine (T). Note how T and C have only a single ring, while A and G have two rings in their molecular structure. Also note that C pairs with G and A pairs with T, to give two pairs of roughly the same molecular dimensions.

"Seeing" DNA

There are many ways to see things. You can see things with the naked eye, with a microscope, a telescope, an electron microscope, or even as computer images that have very little to do with the real appearance of the thing you are interested in. Radio telescopes and DNA sequences fall into this last category. This is because an actual picture of the chemical form of DNA is not that informative. Because our genome is made of DNA, seeing our genome becomes partly a problem of seeing DNA. Again, we can see the fine structure of DNA in two ways. The first is through the "molecular eyes." This way of seeing reveals atoms joined together into four distinct kinds of molecules that are in turn linked to each other, called bases. The other way of seeing is through abstract "DNA eyes." Through these eyes, we see the bases as G's, A's, T's, and C's. [Figure 25]

The ability to "see" DNA is actually the result of some of the greatest scientific accomplishments in human history. As we saw, DNA is a substance that is made of atoms, mostly carbon, nitrogen, phosphorus, oxygen, and hydrogen, that are linked together to make the small building blocks we have called bases. The bases are, in turn, linked together to make long molecules of DNA. This structure makes DNA a real, tangible, biochemical material that can be "seen" at many levels using different techniques. Because it is a chain of chemicals it has dimensions, and if there is enough of it in a single place, we can also see it with the naked eye. What is essential to making genes functional is the order of the bases in the DNA that makes up a gene. Again we are fortunate here, because only four DNA bases—G, A, T, and C—exist in DNA. "Seeing" these bases, or bits of information, and how they are arranged in the DNA sequences of our cells, required technological developments that took about a century to happen.

Seeing DNA happened in stages, with increasing resolution at each stage, and started not with DNA itself but rather with what DNA makes up—genes. We now know that our cells carry our genes. But 100 years ago, inheritance and the way it worked in organisms was a pretty puzzling question to scientists. Darwin had trouble with this very same problem in his own work. While he got evolution, natural selection and variation right, he knew he needed to also have a hereditary mechanism in order for his theory of evolution to be complete. Unfortunately, he missed on the subject of how heredity worked. He came up with a neat but in the end silly idea called pangenesis. But a monk breeding peas, Gregor Mendel, got heredity right at about the same time Darwin was getting evolution right. Mendel worked out the two most important laws in genetics: that genes get randomly mixed up in the sperm and egg (or pollen and ova), and that genes segregate in the egg and sperm. While Darwin wasn't able to "see" genes, Mendel was.

So how do we actually see DNA? What is it we see when we say we have a gene or when we say we have a sequence of a gene? Part of the answer comes from discovering what chemicals genes are made of. By a century ago, scientists had begun breaking open cells, using methods similar to the "Make your own DNA" exercise that follows—and discovered that cells contained two major kinds of chemicals, called proteins and nucleic acids. [Figure 26]

FIGURE 26. A drinking glass with DNA in it produced from the "make your own DNA" experiment in this book. The DNA is the white stringy stuff floating in the glass.

TOOLS AND "CHEMICALS"
Wheat Germ or Yeast
4-inch test tube or short thin drinking glass
Salt
Dishwashing soap
Rubbing alcohol
6-inch glass rod or wooden stick

1) *Take some yeast or wheat germ and put it into a clean test tube. This can also be done in a thin drinking glass. Dried yeast is just clumps of cells from yeast. Wheat germ is just clumps of wheat cells.*

2) *Add a pinch of table salt to the soaked yeast or wheat germ. Mix this in well. What this does is to make the yeast and wheat germ swell. Alternatively you can use salt water and swish it around in your mouth for twenty seconds and then spit the salt water into a tube or slim glass. The salt water will have millions of cells from your mouth in it after you swish.*

3) *Next, add a little liquid dishwashing soap. This step makes the cell and nuclear membranes of the yeast and wheat germ explode. Why? Well, cell membranes are lipids, and soap tends to break up fats and lipids. It's why you wash the greasy pans at Thanksgiving with dishwashing soap.*

4) *Mix this well by swirling. Make sure the soap gets into the yeast and wheat germ solution.*

5) *Finally, very carefully tilt the tube (or thin drinking glass) a little. Take some rubbing alcohol and slowly pour the rubbing alcohol down the side of the tilted tube (or glass). Add about an equal volume of rubbing alcohol as you have of the yeast, salt, soap solution. The rubbing alcohol should form a top layer over the yeast or wheat germ layer. The alcohol makes the DNA, which is hydrophobic, climb up into the alcohol layer.*

6) *Tilt the tube back upright and watch the DNA – the stringy white stuff – float up into the rubbing alcohol layer.*

7) *Very carefully take the glass rod and put it into the white stuff in the isopropanol layer, and twist the glass rod gently. The DNA should stick to the glass rod (or the stick).*

FIGURE 27. Bobble head dolls of Francis Crick and James Watson, co-discoverers of the structure of DNA. Watson and Crick's 1953 *Nature* article is a science classic that described the molecular structure of the double helix. In 1962 Watson and Crick were awarded the Nobel Prize along with Maurice Wilkins. Rosalind Franklin, whose X-ray crystallography was critical in Watson and Crick's work, died of cancer before the Nobel Prize was awarded.

This process is called spooling. The DNA can then be removed from the rubbing alcohol layer and placed on a glass slide.

By the time the 1950s came along, scientists knew that if they wanted to see something really small, they could bounce X-rays off it and collect the diffracted particles from the X-ray on photographic film. If the small things that the scientists were looking at had a regular structure, then the X-ray film would show a distinct pattern of different geometric forms. These techniques are exactly the kind of experiments that the English scientist Rosalind Franklin was so good at. She was known as the best in the X-ray diffraction business in the early part of the 1950s and worked at University of London in Maurice Wilkins' lab, one of the best labs in the world for seeing small things.

At the same time, two of the best puzzle solvers in the history of biology were also working on "seeing" the structure of DNA, but from a more chemical approach. James Watson and Francis Crick were trying to see the structure of DNA at the level of its very basic molecular parts—using their "DNA eyes"—by building models and considering the chemistry involved. [Figure 27] When Watson got a glimpse of one of Franklin's X-rays, he immediately knew that he and Crick were on to something: The patterns on the X-rays looked to Watson just like a double helix! Watson and Crick made their model so that DNA appeared as a double-stranded helix such that one strand of the helix could be paired with the other, using very basic, simple rules of chemistry.

Both Watson and Crick have written at length about the moment of discovery of the structure of DNA, but sometimes it barely seems possible even to put it in words. In fact, their 1953 paper that described the structure of DNA, and hence the hereditary material, was very subdued, but wryly so. Watson and Crick wrote "We wish to suggest a structure for the salt of deoxyribose nucleic acid (DNA). This structure has novel features which are of considerable biological interest...." This quote is only outdone in its understatement by one of the last sentences in the paper: "It has not escaped our notice that the specific pairing we have postulated immediately suggests a possible copying mechanism for the genetic material." Watson and Crick knew from their models that a G (guanine) on one strand of the helix would require a C (cytosine) in the same place on the other strand of the helix. They also knew that an A (adenine) on one strand of the helix would mean that a T (thymine) would have to match it in the same place on the other strand.

We can now see the double helix of a general DNA molecule quite clearly using many techniques; scientists like Watson and Crick were able to see it from physical and chemical data. But this kind of seeing doesn't tell us about genes or the sequence of genes. We also need to see things at other levels of organization of DNA—at the level of how the DNA is packaged in our cells and at the level of the actual order of G's, A's, T's, and C's that make up genes.

"Seeing" Our Chromosomes

To make things easy, we can show that our genomes are subdivided into discrete strings of the four bases that make up DNA—G, A, T, and C, and these are what we call chromosomes. We can see chromosomes under a light microscope, and can easily count them. Different species have different numbers of chromosomes. Bacteria and Archaea, organisms that do not have nuclear membranes, usually have one relatively large, circular chromosome ranging in size from 400,000 bases long to over 10 million bases long. Genomes of eukaryotic organisms that do have nuclear membranes around their chromosomes are highly variable, with chromosome counts of from six in the Indian muntjac to hundreds in some amphibians and plants.

In mammals like ourselves there are two kinds of chromosomes. One specific kind of chromosome in a genome is called a sex chromosome, because it is involved in determining the sex of a mammal. Any chromosome that is not a sex chromosome is called an autosome. We now know that the human genome is made up of 22 "autosomal" chromosome pairs and one pair of sex chromosomes. The autosomes are numbered—1 through 22—in order of size, with chromosome 1 being the largest and chromosome 22 the smallest. Most females have two sex chromosomes that are called X chromosomes, and most males have two kinds of sex chromosomes—one X and one Y.

Oh, and by the way, EVERY normal reproductive cell (sperm if you are a male, and eggs of you are a female) in your body has 22 autosomes and one of the two sex chromosomes, for a total of 3 billion bases. Almost all of your other cells have twice this much in them—44 autosomes and two sex chromosomes, for a total of 46 chromosomes (either two X's and you're a female, or an X and Y and you're a male), giving 6 billion nucleotides in total in cells that aren't sperm or eggs. As we will soon see, this is a lot of DNA for a single cell, and it carries a boatload of information.

As we explained earlier, our chromosomes are made of genes and junk DNA, arrayed a lot like beads on strings that are, in turn, composed of stretches of the four bases of the DNA alphabet. These bases are what make up the code for how our DNA translates to proteins.

To show how much information is in our genome, let's suppose that each of the bases in our genome can be represented by the four letters G, A, T, and C. The three million letters, about the size of a bacterial genome in the same size print as on this page, would stretch about 4 miles. Three billion letters, the number of letters in our genome, in the same size print as on this page (1,000 times 3 million) would stretch 4,000 miles, or all the way across the United States from New York City to Los Angeles and farther. Let's shift the size of the letters to 2-point Courier font (G, A, T, C; the font is really tiny!), instead of their actual miniscule microscopic size, as shown below:

GATCGATCGATCGATC

Using 2-point font, the number of bases on the 23 chromosomes of the human genome could be metaphorically or even physically strung out across the streets of New

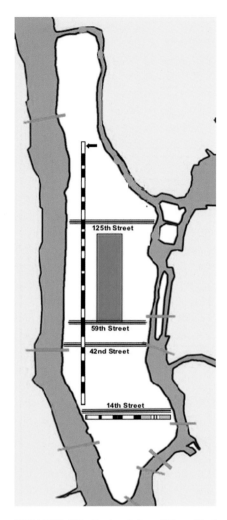

FIGURE 28. Part of the "New York City genome". If each base in the human genome were printed in 2 point Courier font, chromosome 1 (the largest human chromosome at 250,000,000 bases) would stretch from just above 14th Street in Manhattan to about 168th Street. At the very tip of chromosome 1 (arrow) resides the locus for Huntington's Disease. Also at this location in Manhattan is Columbia University Medical School where Nancy Wexler's lab resides. In 2 point Times Roman font the smallest chromosome (at 55,000,000 bases) would stretch along 8th Street from the Hudson River to the East River.

York to give a New York City genome. [Figure 28] In fact, Eric Lander, director of the Whitehead Institute at the Massachusetts Institute of Technology and one of the main players in completing the human genome, suggested putting letters on the streets of New York City to celebrate the completion of the first draft of the human genome. For instance, if 2-point font is used to explain how big chromosome 4 is, the distance that 203 million letters in 2-point font (the number of letters on chromosome 4) would cover, if laid end to end, would be about five miles, or about the distance from 14th Street in downtown Manhattan up to Washington Heights. The smallest chromosome in our genomes, chromosome 22, is about 50 million letters long, and it would stretch along 8th Street from the East River to the Hudson River (8th Street actually turns into St. Mark's Place at 3rd Avenue, and disappears or ends somewhere in the East Village, but it is a good choice for the authors of this book, because we both live near or on 8th Street).

In this same New York City genome, where, exactly, are the genes? How big are they? Let's go to the chromosome 4 that we put on the map from 14th street to Washington Heights. There is a small stretch of the 203 million 2-point font letters, about 10 feet long, that would lie in the Columbia University College of Physicians and Surgeons campus, where Nancy Wexler's laboratory sits at the Columbia Medical School. This happens to be the approximate position on the real chromosome 4 where the gene for Huntington's Disease resides. Dr. Wexler is the person who, during the past two decades, has laid the foundation for our genomic understanding of this terrible disease. So the bottom line with genes is that they are real, they have a defined length, and they reside in defined places on our chromosomes, just as Wexler's lab is real and has a defined location in New York City. Every chromosome in the New York City genome works the same way.

"Seeing" Our Genes

The arrangement of the bases in a gene is what makes one gene do one thing and another gene do another. Early on in the Human Genome Project there was some speculation as to the number of genes in the human genome, which we now calculate is around

25,000. In fact, a website at Cold Spring Harbor Laboratories in Cold Spring Harbor, N.Y., held a contest called GeneSweep to see who could come closest to predicting the actual number of genes in the human genome. Reasonable guesses ranged from 20,000 to 100,000, but the winner was Lee Rowen, a research scientist at the Institute of Systems Biology, who predicted 25,947 genes. Paul Deer of the Medical Research Council Laboratory of Molecular Biology in Cambridge, England, and Olivier Jaillon of Centre National de Séquençage, Paris, France made the next closest guesses at 27, 462 and 26,500 respectively. The prize? Splitting $1,140 and a signed copy of *The Double Helix*.

Most of the entrants in the contest guessed much higher than 25,000 genes. Why? There are two problems that contribute to this guessing game for the number of genes in the human genome. The first problem concerns the large number of proteins—around 100,000—that are needed for the proper development and functioning of a human being. If there were a simple one-to-one correspondence of the number of proteins to the number of genes, the guessing game would have been easy. But a problem arises in using the number of proteins in our bodies to estimate the number of genes. The problem is that many genes make more than a single protein or gene product. The second problem is that not all the DNA in the human genome makes a gene product.

We saw earlier that a genome is not made strictly of DNA that codes for a protein. There are large parts of the human genome that do not code for proteins; in fact, in organisms with cells that have nuclei, like humans, the genes that code for proteins are few and far between. So what is all of this other stuff? To complicate matters even more, genes in organisms with nuclei are interrupted by DNA that doesn't make part of a protein. These are chunks of DNA, called introns, that lie within genes but are not translated into protein. They don't cause cells to make bad or useless proteins, because our cells have figured out ways to process out the introns before a protein is made. In some genes, these introns can be quite large.

When one discounts the parts of the genes that do not make a protein product—the introns—and the DNA between genes that does not make proteins, and estimates the number of bases involved, one sees a remarkable ratio. Some estimates have put the ratio at 1:50, which is to say that only 2 percent of the human genome actually codes for protein.

But why is all that other stuff, that so-called junk DNA, there? Scientists have called it junk DNA because these stretches of DNA in our genome didn't seem to have a job. The so-called junk DNA just seems to sit there and collect dust. When the human genome sequences were analyzed, though, some answers were obtained. Surprisingly, the bulk of the human genome is made up of what are called mobile elements or insertion sequences with straightforward names like SINEs (short interspersed repetitive elements) and LINEs (long interspersed repetitive elements). What these sequences are doing is still kind of a mystery, although it turns out that they are very important to our understanding of relationships among organisms. These sequences might also be important in maintaining the rigid structure of the chromosomes or in regulating gene expression, and might not be so junky after all.

The lack of precision in estimates of the number of genes in the human genome can thus be attributed to how genes are arranged and how the regions between genes are structured, and not to any inaccuracy on the part of scientists. To do any better, scientists had to wait until the entire human genome was sequenced and computer programs had scoured the information for the footprints of genes. The science of hunting for genes in the 3 billion bits of information in the human genome is called bioinformatics. This approach is a brand-new science, made rich by the availability of whole genome sequences for thousands of organisms, especially the human genome.

An OME OME Here and an OME OME There

Now that we know what a genome is, let's see how a genome affects our being human. Scientists have taken to naming the kinds of studies we do in the field of modern genetics with the ending "omics." The field of genomics (GENetics + OMICS) is the area of modern science that generates and interprets the billions of bits of information for genomes of thousands of organisms. Other kinds of studies include transcriptomics, proteomics, phenomics, and even sociogenomics. All of the "omics" mentioned here represent a level of expression of the genetic information we have in the cells in our bodies. This genetic information interacts with other factors and leads to the "construction" of our bodies and indeed to the construction of the cells and bodies of all organisms on Earth.

Phenomics and Sociogenomics

If we go back to the Figure 24 with the DNA sequence and the musician, we see a simple arrow leading from the sequence to the musician. Is this the real story? Actually the arrow between the instructions and the final manifestation of musical ability is way off scale. So much happens in that little arrow and the purpose of much of modern genomics is to understand what goes on there. Scientists interested in this subject work in the field called PHENOMICS (PHENotypes + genOMICS). This incredibly complex area of research is extremely important for a full understanding, for 20/20 vision as it were, of how genomes, gene interactions, and environment influence the final appearance and behavior of organisms. Some scientists have even begun to look for and to "see" genomes and environments in even more complex emergent properties of being human.

The field of "sociogenomics" (SOCIological behavior + GENOMICS) has emerged with the idea that the emergent properties that we see in behaviors and social structures of humans and other organisms can indeed be explained in terms of genomes and how they interact with the environment. Because we can now obtain such large amounts of information from the genome, and because this information can be integrated into studies of behavior, this approach seems viable to some scientists.

Proteomics and Transcriptomics

But how do we go from genomics to phenomics to sociogenomics? There are two "omics" in between—proteomics and transcriptomics—that are critical for a mechanical understanding of the progression from gene to phenomics. From early work in molecular biology, certain rules or dogma about how cells behave were discovered. The most important is called "the **Central Dogma of Molecular Biology**." This dogma states that

DNA **R** RNA ʀ PROTEIN

The statement points out that DNA first makes a copy of itself in a different kind of molecule, called ribonucleic acid or RNA. Then the RNA gets translated into a protein. RNA then becomes an important intermediary in going from the DNA that genes are made of, to proteins, the molecules that do all of the dirty work in the cell. Because both RNA and proteins in the cell can be traced back to the DNA sequence of a gene, they are both called "gene products".

One of the major goals of genomics and phenomics is to assess which genes are expressed or turned on in different tissues. This task is important for many of the human genetic disorders that scientists study. When we want to see if, or how much, a gene is being used, we can look at the amounts of RNA or protein that come from that gene and get a good idea about how active the gene is. This is all because of the central dogma of molecular biology, DNA ʀ RNA ʀ PROTEIN.

To see the importance of getting a hold on how active a gene is, let's look at an example. We will look into human and chimpanzee brains in much more detail in Chapter 8, but for now, let's consider an important question on a lot of scientists' minds. What is it that makes us different from chimpanzees? Of course, the first thing one thinks of is how differently we behave and think. This difference has to be linked to the brain and how it functions. There are many ways to look at the differences between chimp and human brains, but the one approach that gets down to the nitty-gritty of brain structure and function is examination of genes and how they are expressed. To assess which genes are involved, it is necessary to be able to see the gene products of a lot of genes in the brains of chimps and humans, and to see whether different genes are being used in humans' and chimps' brains.

There are two ways to "see" the gene products in the guise of RNA and proteins that interact with each other and the environment to produce the phenome. If you can look at the RNA, that's one, and if you can look at the proteins, that's two. Looking at the RNA is called looking at "transcription," or studying "transcriptomics" (as in TRANSCRIPTion + genOMICS). Looking at proteins is called looking at "translation," or at the genome level, the study of "proteomics" (PROTEins + genOMICS).

More OMICS

One of the most exciting new aspects of modern genomics is the field called proteomics, the study of the proteins that are coded by our genomes. Proteins are a broadly diverse category of molecular structures that do jobs in the cell. Some of the jobs are structural, like holding up the cytoplasm of the cell or holding the chromosomes together. Proteins are a group of large molecules or biochemicals that are made up of smaller molecules called amino acids. These are arranged in a linear fashion, one connected to another. Some proteins have as few as 10 amino acids strung together in a linear arrangement (in a protein called chignolin) or as many as 27,000 amino acids (in a protein appropriately called titin).

Like all other biochemicals, amino acids are made up of atoms, mostly oxygen, hydrogen, carbon, and nitrogen, but also sulfur. There are 20 different amino acids that can make up proteins (some very rare amino acids have recently been discovered to add to the 20 that are primarily used by our cells). What makes one protein different from another and hence makes one protein have a different function from another is the linear sequence in which the different amino acids are arranged in the protein. This linear arrangement is also called the "order of amino acids" in the protein. Protein structures are very intriguing, but more interesting is how proteins interact with one another to produce the complex structures of cells, organs, and organisms. For instance, in the 3-D image of a protein seen in Figure 23, the blue ribbons and strings are the ways protein scientists display the amino acids. Each blue ribbon and string is made of a linear string of amino acids. Any mutation or change that occurs anywhere in these proteins will sometimes result in a change in the shape of the protein. A change in the shape of the protein will often cause a change in its function. So scientists can use these three-dimensional models as predictors of disease and other aspects of human health.

Proteomics also attempts to understand the interaction of proteins, and this process "sees" something completely different. When one uses proteomics to understand protein interactions, it is almost like seeing in more than three dimensions. Protein interactions are predicted by scientists using many techniques. The simplest (easy for us to say) is to test each protein with every other protein for a signal of interaction. For the human genome, where there are 25,000 or so genes, this means that more than 25,000 X 25,000 = 625 million possible pairwise interactions need to be tested. Obviously this is a large number, so shortcuts are being developed to make these tests go faster and to rule out certain pairs that do not need to be tested.

How does the cell make all of these proteins that make up the proteome? From genes, of course, and, as we have seen above, the intermediary between our genes and our proteins is RNA.

And ... Even More OMICS

Transcriptomics is all about identifying and counting the different transcribed genes or RNAs in a cell. It is the difference in transcription of genes that makes a nerve cell different from a blood cell. So the task of transcriptomics is to identify and quantify the genes that are transcribed in cells of different tissues. Sounds overwhelming, but scientists have found a way to do it. When a cell in a tissue or organ goes about its business, it is continually pumping out gene products. Genes are first transcribed into RNA molecules that are copies of the gene, which are then translated into protein. What makes nervous tissue nervous tissue is the kinds of RNA the cells of the tissue are making. So if we can take cells that have a specific function from a tissue, and look at what RNAs are being made in those cells, we can infer which genes are important for making a cell behave a certain way. Looking at the transcriptome has generated one of the most intricate and beautiful tools genomics uses to solve the important problem raised by gathering large amounts of data in single experiments – the microarray. We are now ready to look at our genome at the gene and DNA level.

Reading the Strand of Life

Simply looking at the DNA you can make from wheat germ or from your mouth cells in a tube doesn't do much for our understanding of how genes work or what kinds of information are in genes and genomes. What we really want to know is the arrangement of the bases in genes. It is the arrangement of the G's, A's, T's, and C's in a gene that tells us what the gene will do. This requires that we actually "see" the linear arrangement of the bases in genes.

Because DNA and genes are so small we can't simply look through a microscope and see the arrangement of the bases in a gene, at least not at this point. We say this because, as we write this book, methods are being developed that utilize visualization techniques on single molecules of DNA. For now though, we can say that some very ingenious techniques had to be developed in order to "see" the sequence of bases in a gene. The accomplishments of scientists in their attempts to decipher the human genome in the century since the rediscovery of Mendel's laws are a tour de force of human intelligence and ingenuity. Rather than one simple technique, DNA sequencing is the combination of lots of important techniques developed gradually over the past few decades. But it works, and works well. Getting DNA sequences today is about 10,000 times faster than 20 years ago, and much cheaper. The techniques will only get better. There is even a competition to develop a whole-genome sequencing technique that costs $1,000.

Here is how a genome is sequenced nowadays. There are three main steps to sequencing a genome—preparing the genome for sequencing (Breaking Up), the sequencing itself (Colors), and analysis of the resulting sequences (Puzzling).

Breaking Up Is Hard to Do

The first step is to get the genome or DNA out of the tissues of the organism you want to sequence. Any kind of tissue will do, because almost all tissues have cells with genomes in them. The DNA from the millions of cells of a tissue is separated from the tissue in much the same way we made DNA from the cells of the mouth earlier in this chapter. The DNA isolated this way is really long. For instance, chromosome 1 in the human genome is 250 million bases long, and even the smallest chromosome in our genome—chromosome 22—is better than 50 million bases long. DNA molecules this long are, simply put, a pain for scientists to manipulate. Modern sequencing techniques work best with fragments ranging between tens of bases and thousands of bases.

To get the right size DNA to work with, ultrasound is used to shear the DNA into fragments of about 2,000 bases. At very high pitches of sound, DNA destabilizes and breaks in random places, but it does so very regularly with respect to distance between the breaks. These fragments will be DNA pieces sheared at random spots in the genome, but all have similar sizes. Depending on the pitch of the sound, you can make longer or shorter pieces of DNA. Because your genome preparation has DNA from millions of cells, the random shearing ensures that no two fragments of DNA in your sheared DNA are the same. In other words, there are parts of any given sheared fragment in at least a few of the other fragments in the genome preparation. The sheared DNA pieces are then placed into DNA holders called plasmids, using biochemical methods. The sheared DNA in the DNA holder or plasmid is called an insert. This step makes lots and lots of the insert, of a manageable size, available for subsequent steps.

Colors of the Rainbow

The next step is to treat the DNA holders plus inserts with several biochemicals that have four different fluorescent dyes in them. This biochemical reaction has one dye for each of the four bases in DNA. The fluorescent dyes that are used are different from each other in that they will fluoresce at different wavelengths when laser beams are pointed at them. When they fluoresce, they give off a signal that can be collected by a detector device. For convenience, these signals are represented in subsequent steps by four colors: usually red is a T, green is an A, black is a G, and yellow is a C. The chemical reaction that ensues simulates making a DNA strand with the specific fluorescences for the four bases attached at the right positions in the insert DNA. The chemical reactions are then placed on an automated sequencer that simulates stretching the insert DNA out like a long string so the fluorescently labeled bases can be detected by a tiny highly focused laser beam.

Puzzling

A computer collects all the fluorescence data and converts it to a readout that reflects the actual sequence in the insert DNA. For a genome the size of the human genome, millions of DNA holders plus inserts are treated. For a sequence the size of a bacterial genome tens of thousands of DNA holders plus inserts are treated. The size of the genome dictates how much work one needs to do to get the whole genome of an organism. All of the sequences so generated now need to be put back together again, like a puzzle. Because the inserts were generated randomly, there should be overlap on the ends of the insert. These overlaps act like interlocking puzzle pieces. Because most humans have trouble with even 1,000-piece puzzles, a computer is used to do this last step, which is called assembly. The next and final step is a process whereby the assembled DNA sequences are scanned for genes and the genes are given names. This last step is called annotation. Voila! You have a genome.

Whose Genome Is on That iPod?

There is much lore about who exactly it was who suggested that the human genome could be sequenced. By most accounts though, the suggestion that a human genome project could be a reality was made in 1985 at a meeting convened by Robert Sinsheimer, then chancellor of the University of California, Santa Cruz, California. Following the initial suggestion came more fantastic predictions about having the genome sequence of a human. Walter Gilbert, then of Harvard University and always on the cutting edge of molecular biology, made the suggestion that one day soon everyone would be walking around with the sequence of his genome on a card the size of a credit card. Gilbert suggested that this would be a useful tool for personalized medical treatment. You go to the doctor with an ailment, even a common cold, and the doctor scans your genome card and determines that the best course of treatment for your cold is a specific kind of antibiotic treatment that is best suited to your own immune system.

It turns out that Gilbert's prediction was right, and just about on schedule. Scientists working on the results of the human genome can cram all of the information from a human genome (3 billion bases or bits of information) onto a 5 gigabyte iPod. Not quite a card, but certainly small enough to fit in your pocket. Technology will, sooner rather than later, easily proceed to the point where the information can be loaded onto a card.

In addition, there was much controversy about the prospects of getting the entire sequence of a human being. Many biologists asked, why bother? As we saw in the first

FIGURE 29. Who's DNA is in that genome database? Portraits of several famous geneticists whose work has been critical in our understanding of genetics and genomics. From left to right – Charles Darwin, Francis Collins, Craig Venter, Rosalind Franklin, Barbara MacClintock, James Watson, Gregor Mendel and Franz Meischer. Collins and Venter led the public and Celera efforts respectively that sequenced the human genome. Franklin and Watson were instrumental in the discovery of the structure of DNA. Mendel of course was the father of modern genetics and if he was the father then MacClintock is the mother of modern genetics. Meischer was the first person to discover that DNA existed.

chapter of this book, science is structured around hypothesis testing, where we make statements that we can falsify through experimentation or through collecting information. The problem with sequencing the human genome that a lot of scientists saw in the 1980s, when the initial suggestion was made, was that they didn't see the sequencing project as anything more than "sequence first, ask questions later."

Additionally, the question of whose genome to sequence was problematic to most biologists, because a single genome is just that—a single data point. What about the 5 billion or so other human genomes that existed on the planet in 1986? On the way to figuring out whose genome to sequence, some scientists, tongue in cheek, suggested exhuming Darwin and using tissue from his corpse as a source of the DNA to be sequenced. [Figure 29]

The two groups that presented first drafts of the human genome – one an international coalition and the other a company (Celera) – used basically the same approach. They both settled on mixing the DNA of anonymous donors from very different ethnic backgrounds and then using this mixture to obtain the sequence. Celera had its own genomic cocktail, as did the international effort. One big difference between the two cocktails, though, was that the Celera cocktail had the boss's DNA in it too – not Bruce Springsteen but rather Craig Venter, president of Celera during the human genome period, who recently admitted to having his DNA put into the Celera genome cocktail. In another ironic twist, it turns out, as a feasibility study, a biotech company, hoping to show the utility of their sequencing method, has sequenced the genome of James Watson (third from the right).

Why is This All Important **to Human Existence and Origins?**

Getting the entire sequence of G's, A's, T's, and C's of the whole human genome gives us a shot at understanding how we can go from a set of written instructions in the form of chemical structures (the bases) to the complex organisms we see around us. Genomes have become central to how we approach the subject of our makeup, not because they are the most important aspect of our existence, but because they are a point of reference for hard data. Genomes are a great starting point for understanding life. The whole human genome sequence results in 3 billion bits of hard information in the guise of 3 billion bases. The bits of information are there, they are real, and they are used by our bodies to jump-start and guide what we are. The sequence of the human genome and all of the work being done around this project has also been suggested to be one of the great starting points for advancing human health. If we know the DNA sequence of a gene that produces a disease in a human genome, then we can at least start to understand the disease, diagnose it, and begin to look for cures for it.

How organisms are put together, how they work in a molecular genetic sense, and how they react to the environment are critical to our understanding of human origins. By comparing the genomes of primates, other mammals, and other animals with the way we are "constructed," we can learn about the differences in our appearance and behavior from those of our close relatives. These differences in construction and behavior can either be the cause of our uniqueness with respect to other organisms or the result of being separated from other organisms during the evolutionary process. The former is at the heart of trying to find the genetic basis of our unique appearance and behavior (Chapter 8) and the latter allows us to use DNA sequences to understand our history (Chapter 7).

The footprints of divergence and the history of how we evolved and are evolving are also hidden in the sequence of the bases in our genomes. All organisms have genomes, and all genomes from organisms living and dead on this planet emanate from a single common ancestor. Because changes in DNA are passed on from parent to offspring, there is a continual record of the changes that have occurred over the 3 billion years of the existence of life on this planet. So the genome is a logical stepping-off point for almost all of the recent exploration of our biological and historical being. The following chapters will delve into how our genomes and bones collude to give us an excellent picture of our humanness.

CHAPTER FOUR

EVOLUTION AND HUMAN ORIGINS

FIGURE 32. Vial of poly U from Nirenberg's experiments, alongside one of his lab notes.

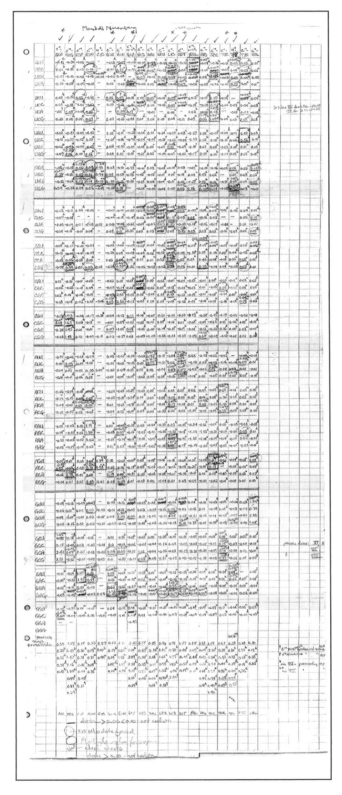

codes for P (change the C's in RNA to G's in DNA). Next, when the following string of nucleotides (ACC)$_n$ (ACC's repeated), was synthesized, an amino acid string of prolines was made. This experiment clearly showed that the triplet GGT codes for P. That might be taken to show that ACC in RNA is made from GGT in DNA—but doesn't ACC in RNA come from TGG in DNA? Actually, no. Remember that when a nucleotide chain is synthesized, it is done from a template strand of DNA that is running in the opposite direction. Thus, the RNA nucleotide chain of 5ACCACCACCACC3 is synthesized from the DNA nucleotide chain of 5GGTGGTGGTGGT3. This is better visualized as follows:

    5'ACCACCACCACC3'   RNA product
    3'TGGTGGTGGTGG5'   DNA template

This second string of P's, made by the triplet GGT, showed that the genetic code is redundant. Indeed, when the RNA triplets (TTC)$_n$ and (GCC)$_n$ were synthesized and tested, they too gave strings of amino acids with P's in them, indicating that the amino acid proline is coded for by four triplets: GGG, GGA, GGT and GGC. Note that it is the third position in the triplets that seems to be the one that can change. When any other nucleotide is put in either the first or second position of the triplet, a chain of amino acids is produced that is composed of something other than Ps. This observation pinned down the fact that the third position can usually wobble about a bit and you can still get the same amino acid. So, after testing all 64 possible combinations of G,A,U and C's, scientists were able to decipher the genetic code. All 20 amino acids are accommodated by the code, and there are even three triplets that tell a protein where it ends. These three punctuation marks were dubbed "terminators" (way before Arnold Schwarzenegger was an actor let alone a governor), and they act like periods at the end of a sentence. More work determined that one amino acid, methionine, or M, was at the beginning of almost every protein. There is a single triplet that codes for M (ATG); it is called the initiation codon, and acts like the capitalization at the beginning of a sentence.

So not only does the genetic code allow translating the RNA sequence into protein, but it also allows for the punctuation (capitalization indicating the start of the protein, and a period indicating the end of a protein). If we know what these 64 triplets mean, we can now look at any DNA or RNA sequence and determine not only where it occurs in the genome, but which protein any part of the genome codes for. In fact, one of the first steps in genomics is to look for ATG triplets and for the three terminators. Some of the positions in a triplet, most notably the third, can be what scientists call "silent" and can be changed with no effect on what amino acid will be placed in the part of the protein the triplet designates.

And now that we have the genetic code and mutation under our belts, we can go on to see how DNA sequences and paleontological as well as anatomical information can be used to examine both descent with modification and changes in genes over time at the population level.

Natural Selection, **Move Over**

One of the many things Darwin did right was to offer a mechanism for how evolution occurs—natural selection. But natural selection is not the only way for evolution to work, for natural selection is not the only force in nature that can cause change in genomes over time. There are, in fact, five major ways genomes in populations can change with time—natural selection, migration, genetic drift, altered mating patterns such as inbreeding, and mutation. Of these five, only two are particularly effective at changing allele frequencies with time. One of them, mutation, happens so infrequently that, by itself, it will have very little effect on allele frequencies. Mating system alteration and migration may cause local changes in allele frequencies, but they don't really alter the genetic structure of species on any large scale.

So let's look closely at the remaining two mechanisms that are the ones that seem to do the hard work in changing allele frequencies over time—drift and natural selection. Genetic drift is a phenomenon that occurs because of what a statistician would call sampling error. This kind of error is really just the luck of the draw. As we've already seen, when you are dealing with large numbers there is little luck to the draw. If you flipped a fair silver dollar 1,000 times, you would almost always win the bet that you would get about 500 heads and 500 tails, give or take 10 or 20. But if you flipped the coin four times, the bet that you would get two heads and two tails wouldn't always win. In fact, there is probably no bet on the frequency of heads and tails that would be a sure one. On the other hand, if you bet someone that in 20 tries you could flip four heads (or tails) in a row, you might be in luck.

While this bit of information might make (or lose) you a little cash sooner or later, it also demonstrates that larger samples perform better with respect to obtaining expected outcomes based on frequencies. It also shows that smaller samples will perform very poorly at giving expected frequencies. Now, substitute the word "population" in the previous two sentences for the word "samples" and you have the essence of genetic drift, though instead of flipping coins, your expectations are now about the frequencies of genes.

To visualize this analogy, consider a population of 500 organisms having a gene with two forms or alleles that we will name "big A" and "little a." Big "A" and little "a" are both forms of the same gene, but they have different DNA sequences. Let's say that big "A" has a frequency in the population of 50 percent, and little "a" has a frequency of 50 percent. Because, as Mendel showed, alleles for a gene are normally passed on randomly to the next generation (this process involves "random assortment" and "segregation"), if we have a relatively large population of organisms the expectations for allele frequencies in the next generation will be 50 percent big "A" and 50 percent little "a." In fact, if there are no outside perturbations like selection, inbreeding, migration, or massive mutation, this population of 500 individuals will *always* have about 50 percent big "A" and about 50 percent little "a"—forever! In other words, this population will not evolve.

Now, consider a starting population of four individuals with 50 percent big "A" and 50 percent little "a." The probability that this population will produce offspring with 100 percent big "A" is higher than in the 500–individual case. But the probability that this population will produce offspring with 100 percent little "a" is also higher than in the 500 individual case. All of this means that there is a higher probability that one or the other allele will be eliminated in a smaller population. This loss of an allele can happen even if the allele that gets eliminated is a "better" or more fit allele than the one that gets fixed.

There are two related phenomena with respect to drift. One, known as "bottlenecking," occurs when a once-large population has become very small. The bottleneck effect will create the conditions necessary for genetic drift, and if the population stays very small after the initial bottleneck its effects will be even more pronounced, resulting in the random fixation of some alleles. An especially severe kind of bottlenecking is called the "founder effect," wherein one or a few pregnant females start a new population unconnected to the parent population. Where it occurs, the founder effect may have a very profound effect on the fixation of alleles in natural populations. Let's look at a few examples.

What do Gerrit Jansz, Arnold, Samuel King, William Machado, Patient 19, the Residents of Pingalap, and Altagracia Carrasco have in common?

All were involved in human bottleneck events that produced genetic anomalies in high frequency. Along with his wife, Gerrit Jansz immigrated to South Africa in the 1600s. Jansz was afflicted with a metabolic problem called porphyria. This disorder causes mild pain in the chest and limbs due to the production of malfunctioning heme (oxygen-carrying) molecules in those affected. Three hundred years after the Jansz family moved to South Africa, more than 30,000 people now have the disorder, and all can trace their genealogies back to Gerrit Jansz. Another South African founder was a Chinese man named Arnold, one of a small number of original Chinese immigrants who gave rise to the current population of 100,000 South Africans of Chinese ancestry. Arnold is gone, but he has nearly 400 descendents as a result of having had seven wives. Nearly one-fourth of his descendants have the dental disease that he carried with him to South Africa.

Samuel King moved to the Pennsylvania Amish region of the United States in the 1700s. He carried an allele that was, two centuries later, named Ellis-van Creveld Syndrome allele (after the two doctors—Richard Ellis of Edinburgh and Simon van Creveld of Amsterdam—who first described it in the 1940s). This syndrome causes short-limb dwarfism, polydactylism (additional fingers or toes), malformation of the bones of the wrist, dystrophy of the fingernails, partial hare-lip, cardiac malformation, and often pre-natal eruption of the teeth. King himself probably did not have the disorder, but rather carried a single copy (allele) of it to his new home in Lancaster County. Because he was a founder of the Amish settlement and had many offspring, the frequency of the Ellis-van Creveld allele in subsequent generations was very high due to inbreeding. And this, of course, increased the probability that some offspring of his descendents would get a copy of it both from their mother and from their father—and thus be affected by the syndrome, which is found in a much higher frequency in this Amish population than in other populations in the world.

When the Portuguese settled the Azores, the founding population was a small group. Within the last century, small numbers of people emigrated from the Azores to the northeastern United States. Among this group was a man named William Machado, who was affected by a discomforting form of ataxia. This syndrome causes atrophy of the muscles and diabetes mellitus, and it is now found very commonly in those Massachusetts towns where the Azorean immigrants landed.

In a similar vein, Patient 19 was an anonymous Northern European who had Sjogren-Larsson syndrome, a devastating genetic disorder also known as spastic quadriplegia. Patient 19 was part of a large pedigree with its ancestry in northern Sweden. Torsten Sjogren and Tage Larsson, the Swedish physicians who characterized the disorder in 1957,

FIGURE 33. On right is a scene from a small Pacific Island. On the left is how a Pingalap Islander with color blindness would see the same scene. Courtesy genomenews.org.

showed that the affected patients have two copies of the defective allele, and that this allele became predominant in a founder event that occurred about 600 years ago in Sweden. Today, about 1.3 percent of the people of northern Sweden carry one copy of the gene in their genome.

In his book *The Island of the Colorblind*, neurologist Oliver Sacks described the visual and psychological effects of an interesting situation on a small South Pacific island. [Figure 33] Since he published the book, the genetic basis for the colorblindness of people from this island has been worked out using genomic techniques. The residents of Pingalap have an unusually high frequency of total colorblindness. This visual anomaly is the result of a mutated gene on the second-largest chromosome in our genomes, Chromosome 2. Specifically, this change affects the gene CNGB, which is essential for the proper function of cone cells, the color-detecting receptors in our eyes that we encountered in our voyage to the center of the cell. The discovery of the genetic cause of the colorblindness is interesting in and of itself, but the reason for the high incidence of this form of colorblindness in the islanders of Pingalap is even more fascinating. About 200 years ago, in 1775, Typhoon Lengkieki devastated the island. The population was decimated by the disaster and the ensuing famine. The handful of survivors left after this bottleneck repopulated the island, and just one of them had the gene for colorblindness. Yet today, of the 3,000 people living on the island, about 150 are completely colorblind.

Finally, Altagracia Carrasco was a female founder of a small village in the mountainous hinterland of the Dominican Republic. She carried an allele for a defective enzyme called steroid 5 alpha reductase 2. This enzyme is important in the conversion of testosterone to dihydrotestosterone (DHT) in the developing male embryo. DHT, in turn, is important for the determination of the external genitalia in the developing male embryo. Testosterone can do this too, but much larger amounts of it are needed than DHT. So what happens when a male embryo has two copies of Carrasco's allele? Because there is no DHT and not enough testosterone, the male embryo does not develop a penis and hence is born with apparently female genitalia, and with undescended testicles—a pseudohermaphrodite. The story isn't over there, though. As the young pseudohermaphrodite develops through childhood and reaches puberty, a rush of testosterone characteristic of adolescence happens (as any parent of a son going through puberty knows). This time, enough testosterone is produced for the genital area to begin to develop a penis. The syndrome is locally known as "guevodoces," which loosely translates into "eggs at 12" or more appropriately (or inappropriately) "balls at 12." Altagracia Carrasco's genetic legacy still exists in some Dominican Republic populations, to the extent that social rites exist that acknowledge the apparent sex change. One last note is that many different genetic lesions result in pseudohermaphroditism. Some of these involve other enzymes in the conversion of testosterone to DHT, while others affect the very same steroid 5 alpha reductase 2 discussed above. Surprisingly, it turns out that in the Dominican Republic alone, two or three different kinds of altered 5 alpha reductase 2 alleles cause pseudohermaphroditism.

One would expect that, if a bottleneck occurs, many genes would be affected. This scenario is exactly mirrored in the populations of the Eastern European Jewish people known as Ashkenazim. Historical records indicate that a bottleneck occurred in this population around 75 A.D., at the beginning of the Jewish Diaspora, followed by even more severe bottlenecks in 1100 A.D. and 1400 A.D. These bottlenecks resulted in very high frequencies for several metabolic diseases involving a function called lysosomal storage. These include Niemann-Pick Disease, Tay-Sachs Disease, and Gaucher disease, plus other devastating genetic disorders such as the neurological ailment known as Canavan's disorder, and, perhaps the most devastating disease in the whole group, breast cancer caused by the breast cancer 1 gene, or BRCA1.

Anomalously high frequencies of particular genes suggest that many other groups of people have experienced similar bottlenecks. But care is needed when you assume a founder event, and particularly when you interpret the genetics of populations believed to have undergone founder events. For instance, the population of Iceland was thought to have been established by a very small number of founders. Indeed, it was thought by some that just a couple of Vikings were involved. One genome analysis company, deCODE, decided to take advantage of the idea that Icelanders were established by a small founder population and were subsequently isolated for 1,100 years. This supposed small founder population, coupled with the millennium-long isolation from other Europeans, led the scientists at deCODE to suppose that Iceland has a homogeneous genetic composition and that the alleles causing disorders such as multiple sclerosis, schizophrenia, and stroke are the very same alleles that the mythical founder Vikings brought to Iceland. Because there is excellent pedigree information available for most of the families on Iceland, the businesspeople at deCODE believed that Icelanders would be an excellent group for studying and possibly finding cures for those genetic disorders that are found in high frequency on the island.

But when scientists politically opposed to the deCODE group looked at how variable Icelanders' genomes are compared with other Europeans, they arrived at the surprising

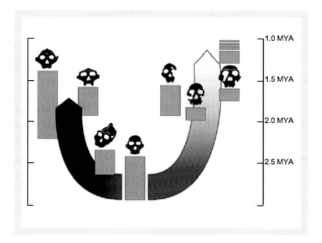

FIGURE 34. The role of selection and genetic drift in the evolution of hominids. The arrows represent evolutionary paths where the darker the arrow, the more important natural selection is believed to have been. Note that the lineage leading to the robust australopiths requires strong natural selection whereas the path to *Homo* is lighter and hence involves more genetic drift. Redrawn from Rebecca Ackermann and James Cheverud, "Detecting genetic drift versus selection in human evolution," *Proceedings of the National Academy of Sciences*, 101-17946-17951.

result that Icelanders are as variable as the French—if not more so! The scientists at deCODE now appear to have backed away from the original claim that the entire Icelandic population is very homogenous, and have instead adopted the notion that isolated pockets of Icelanders, and genetic information from these smaller groups of people, can be more useful in disease-gene hunting.

This turns out not to be surprising, for two reasons. For a start, a more detailed look into the history of migration to Iceland revealed that the founder population probably consisted of thousands of individuals, not all of them of Viking descent. As many as 20 to 40 percent of the founders in Iceland were of Celtic stock from the British Isles, while the rest were of Nordic origin. Further, because sexually reproducing organisms receive one genome from the mother and another from the father, we tend to be quite variable in our own genomes. And finally, scientists using mathematical techniques to study the theory of bottlenecking showed clearly in the 1970s that even founder populations of a very few individuals will not significantly reduce the heterogeneity of a population for *most* genes in the nuclear genome.

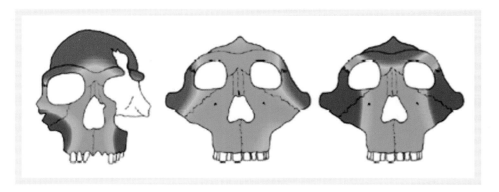

FIGURE 35. Where selection pressure would need to act in order to produce the transition from a gracile australopith to *Homo* (left), a robust australopith from a gracile one (middle) and a late robust australopith from an earlier one. The colors represent the relative instensity of selection pressure where red is strong positive selection, green is no selection and blue is strong negative or purifying selection. (see figure 34.)

The population drift effects we have so far discussed are not just relevant to modern human populations [Figure 34]. It is pretty obvious from the examples given here that these mechanisms have been very important in the recent evolution of genes and genomes in human populations, but what about populations of other species in the genus *Homo*, and of our close primate relatives? Scientists have now begun to examine drift and selection in various hominid primates. [Figure 35] For example, by looking at facial structure of ancient hominid cranial specimens, and by measuring eight aspects of the shape of these fossils skulls, scientists were able to ask whether the changes they observed were the result of random or selected change. They concluded that while early divergence of australopiths was the result of positive natural selection, in contrast, the more recent divergence of species within the genus *Homo* has resulted from random drift.

Naturally, It's Selection

We have already alluded to the "Modern Synthesis," a coherent view of the evolutionary process that took hold during the 1930s and 1940s and that by mid-century had more or less established a stranglehold on evolutionary thought. Essentially, the Modern Synthesis reduced a huge diversity of evolutionary phenomena to a single central notion of gradual changes in gene frequencies within populations under the guiding hand of natural selection. In this way, natural selection became more or less the be-all and end-all of evolution. As Darwin proposed it, natural selection seems blindingly obvious—which is why Darwin's foremost early defender, Thomas Henry Huxley, famously declared, "Why didn't I think of that?" after hearing of it.

The basic idea is that in virtually all organisms, more offspring are produced than ever survive to attain maturity and, more importantly, to reproduce. All populations consist of individuals who vary in their physical and behavioral characteristics, most of which are, of course, heritable; and the ones who succeed in the reproductive stakes will be those that are "fitter," or better adapted to prevailing circumstances, and who will preferentially transmit their superior adaptations to their offspring. In this way, the alleles underlying those favorable adaptations will be disproportionally passed along, and their frequencies will increase from generation to generation. At the same time, alleles that do not favor successful reproduction will tend toward elimination from the gene pool. Over time populations will thus evolve, as their gene pools are modified.

The appeal of natural selection as *the* mechanism of evolution is manifest in the way the Modern Synthesis took over evolutionary biology. The concept of natural selection is not only easy to grasp, but it seems to appeal particularly to the human reductionist propensity. What's more, it is easy to find cases demonstrating that selection is actually taking place. One of the most striking modern examples of selection in natural populations is the development of drug-resistance by the HIV viruses that cause AIDS in humans. [Figure 36] HIV is a devastating infectious virus and has a high mutation rate because its genome is made of RNA, and it turns out that RNA genomes mutate with great ease. The resulting high mutation rate means that even within a single infected individual, there is a great deal of variation in the virus. When a human infected with HIV is treated with an anti-viral drug, the immediate effect will be to reduce the number of viruses present. Because so much variation exists in the HIV

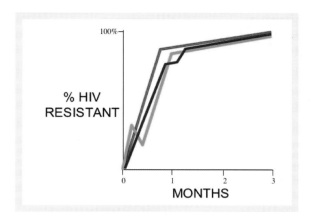

FIGURE 36. Graph of the percent of viruses in three patients (red yellow and blue lines) that are resistant to an antiviral treatment, versus time. Note that all three patients accumulate more and more resistant viruses from the time of first treatment until almost all viruses are resistant. This is literally a text book example of natural selection in action. Redrawn from Neil A. Campbell and Jane B. Reece, *Biology* (2004), Benjamin Cummings.

population, however, some small number of the HIVs will be resistant to the anti-viral treatment. It doesn't take long at all for those resistant HIVs to replicate into large numbers—and before long, the treated individual will be infected with a large population of resistant HIV. In the example in Figure 36, the HIVs are the variable population, and the anti-viral is the selection agent. The viruses that are resistant are more "fit" than those that are susceptible, and after a short period of time, more of the fitter, resistant viruses will dominate.

One question that arose after drift was recognized as a potent force in molecular evolution was whether it is selection or drift that is the major evolutionary force. This controversy was actually a major focus of population genetics when molecular genetic techniques first started to be used in evolutionary biology in the 1960s and 1970s, and it became known as the selection neutrality debate, because from a selective point of view genes whose frequencies change through drift should be "neutral." Now that we have whole genome sequences and the ability to sequence large parts of genomes of organisms, this selection neutrality question has been addressed at length.

In essence, there are three major ways of looking at the world with respect to neutrality and selection. In one world, everything is selected. In this world, whether selection is against (deleterious) or for (advantageous), there is little room for neutrality. In another possible world, the majority of variants are neutral, with only a very small proportion of the variants being selected. Because many scientists doubted that there could be as much neutrality as this middle world implies, they hypothesized yet a third world, in which selection occurs but a large majority of the variants are nearly neutral. Which of these three possible worlds accords with the world we actually see? Many feel that the third possibility is the most reasonable description of what actually happens within populations, leaving room for selective evolution as well as neutral drift. But only more genome information will assist in telling us whether selection or drift is at work in the majority of cases. As a result, much effort in human population genomics is now directed toward detecting natural selection at the molecular level.

We could go on and on with examples of natural selection in action, or dive into theoretical mathematics, showing the appropriateness of assuming natural selection in natural populations, or use genome sequences to demonstrate the amount and degree of natural selection at the genome level. But you would be hard-pressed to find an evolutionary biologist who does not accept that natural selection is a significant factor in shaping the frequency of alleles in natural populations. Given this, we want to delve into two

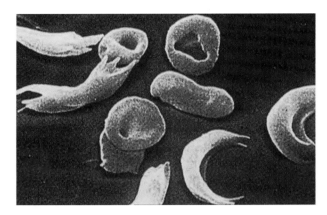

FIGURE 37. Sickle shaped cells in the blood of a person with two copies of the sickle cell allele.

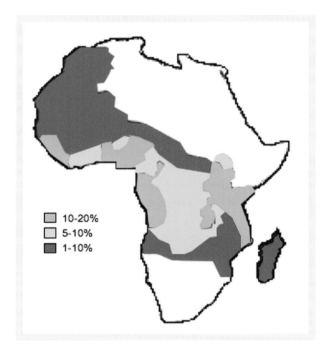

10-20%
5-10%
1-10%

FIGURE 38. Map of sickle-cell allele distribution in populations of *Homo sapiens* in Africa. The bluer the color the higher the frequency of the sickle cell allele.

aspects of natural selection that are counterintuitive to the ideas that most people have about natural selection. The first of these suggests that natural selection does *not* necessarily always involve the survival—or, rather, reproductive success—of the fittest. In fact, natural selection can sometimes prevent change of allele frequencies to favor the fittest. The second addresses the question of whether evolution is directional, and whether the changes over time that we see in the fossil record are specifically adaptive, rather than drift-related or secondary to other changes.

In most simple cases, it can be shown that evolution is a straightforward process where the fittest win. For instance, one of the most shopworn examples of natural selection and adaptation in human populations is the case of sickle cell anemia, a syndrome that is found in high frequencies among people living in Africa and in people of African descent. [Figure 38] This anemia is caused by a simple mutation in the ß-globin gene of humans. In the triplet for the sixth amino acid in the ß-globin protein a nucleotide substitution has occurred, changing the normal glutamic acid to an alternative called valine. This change in amino acid causes a change in the shape of the protein, which is found in our red blood cells. The change in the ß-globin protein causes the chemistry of red blood cells carrying the sickle cell allele to go awry. ß-globin is part of hemoglobin, which is the oxygen-carrying component of the red blood cells. If the body doesn't produce enough hemoglobin, the red blood cells will distort and fail to carry the proper amount of oxygen to the tissues of the body. Thus people who have two copies of this allele in their genomes, one from each parent, have red blood cells that deform into "sickle" shapes. This sickling causes an onslaught of physiological and neurological problems as a result of the reduced oxygen supply to the tissues.

In Africa, where the sickle-cell allele (S allele) originated, it is found in very high proportions both in individuals with two copies of S in their genomes, and in people who carry only one copy of S. [Figure 38] Some blood anemias of this kind caused by a single defective allele are called thalassemias, and they happen to be common where mosquitoes that transmit malaria are found. It turns out that in sickle cell disease this happens because having one S allele confers resistance to malaria, which is why in Africa the frequency of the S allele is higher in areas where malaria is more prevalent. The other allele (the A allele) will never cause sickling, whether it is present in one copy or two. So we have the following situations:

In environments without malaria:

A + A does not sickle; resistance to malaria doesn't matter.
A + S does not sickle; resistance to malaria doesn't matter.
S + S sickles; resistance to malaria doesn't matter, but very ill.

In malarial environments:

A + A does not sickle; not resistant to malaria: a problem.
A + S does not sickle; resistant to malaria!
S + S sickles; so ill that resistance to malaria hardly matters.

Just by looking at these descriptions, one can conclude that in the environment without malaria the A+A and A+S individuals will be the fittest, because sickling is the limiting factor. In this kind of population, S will eventually be lost because of the low fitness of S+S individuals. But in the malarial environment, the A+S individuals will be most fit because malaria is the limiting factor. A+A individuals will be susceptible to malaria, while S persists in such populations because it is favored in the A+S individuals. The sad byproduct of maintaining S in the population to maintain A+S individuals, who are resistant to malaria, is that chronically ill S+S individuals will also be produced. These instances, where individuals such as A+S are fittest, is called heterozygote advantage.

The story isn't over yet, though, because there are many forms of anemia alleles in the world. In western Africa, there is another sickle cell allele found in low frequency that is called C. C is a "super allele," in that when individuals have the C+C combination in malarial environments, their red blood cells don't sickle and they are highly resistant to malaria. Then why aren't all people in malarial areas C+C? Because C coexists in the same gene pool with A and S, and some strange things can happen. The C allele in a population with A and S will result in A+C and S+C individuals, too. So in addition to A+A, A+S, and S+S, we have to now add C+C, A+C, and S+C individuals to the picture:

A + A does not sickle; no resistance.
A + S does not sickle; resistant.
S + S sickles.
C + C does not sickle; super-resistant.
A + C does not sickle; no resistace.
S + C sickles; resistant.

We can arrange these allele combinations, when in a malarial environment, in the order of their fitness, where the arrows indicate decreasing advantage:

```
C+C > A+S > A+C, A+A > S+C, S+S
```

This arrangement says some interesting things about natural selection. The reason that C+C will never take over this population (assuming the size of the population stays large) can be seen by looking at the position of the S+C and the A+S individuals in the ordering of these combinations. The only way to bump the population to C+C would be for a severe bottleneck to occur where the S allele was eliminated: always a possibility, but an essentially random result rather than one related to natural selection.

The sickle cell story is relatively simple in the context of adaptation, because it concerns only a single gene—ß-globin. But usually things are more complicated than this. Most traits that are important in increasing fitness are reproductive traits, things like the ability to produce lots of offspring, age at first reproduction, and other traits that are complex in their genetic architecture.

Other traits that affect the health of individuals are also complex. For instance, asthma is considered a complex trait. The disease we consider asthma is probably more than a single phenomenon, meaning that there is more than one way to have asthma at the genetic level. Asthma is controlled by several genes in the human genome that scientists are just now deciphering. Chromosomes 5, 6, 11, 12, and 14 all have genes on them that are involved in asthma. Chromosome 5 seems to be a hot spot for asthma genes, though, and much effort has been spent on this chromosome to locate the most important genes involved in the disease. Heart disease, mental disorders and cancers are all very similar to asthma, in that they are complex at the genetic level.

Do these complex traits and diseases follow the same natural selection rules as sickle cell anemia? Yes! The analysis of these complex traits becomes more difficult and involved, though the individual genes stick to the same rules, often with very similar results – the fittest alleles often go nowhere. The interactions of the many genes involved in a complex trait such as asthma have to be considered when predicting the behavior of these complex traits.

Calvinism

The late Stephen Jay Gould and the living Richard Lewontin, Alexander Agassiz Research Professor at the Museum of Comparative Zoology, Harvard University, once characterized the prevailing view of evolutionary biology just after the Modern Synthesis as "adaptationist." They were worried that too many "just-so" stories were being concocted to explain patterns we see in nature. They called this approach the Panglossian Paradigm, after a character in Voltaire's *Candide*. We prefer to call this problem the Calvin Effect, not after John Calvin, the 16th century theologian and reformer, but rather Calvin of Calvin and Hobbes.

To the wiser-than-his-years little boy in this comic strip, everything seems to have a purpose. [Figure 39] In one vignette, Calvin suggests this about his wardrobe: "What's the point of wearing your lucky rocketship underpants if no one asks to see 'em?" As

FIGURE 39. Calvinism!

for dreams, Calvin has this to say: "I think we dream so we don't have to be apart so long. If we're in each other's dreams, we can play together all night!" To Calvin, lucky rocketship underwear is to be worn for people to admire, and dreams exist so he can play all night. These would be plausible explanations for dreams and underwear in a child's world, but as we know, there are better reasons for the existence of underwear and dreams. Perhaps the most relevant example of the Calvin effect, with respect to the later chapters of this book, is embodied in the following Calvin quote: "Is it man's purpose on Earth to express himself, to bring form to thought, and to discover meaning in experience? Or is it just something to do when he is bored?" Here once again, Calvin has a unique hold on the metaphysical, and again offers a plausible mechanism for the origin of creativity, but one that is adaptationist.

In many ways, this quote and the comic strip itself embody our main point about evolutionary mechanisms and the way scientists approach understanding these mechanisms. Calvin can come up with an explanation for everything. Most of the explanations are self-serving, or used to evade losing his dessert rights. But one thing's for sure, Calvin is the modern-day Dr. Pangloss.

The good thing about the Panglossian Paradigm (the Calvin effect) is that imagination can run wild, and very interesting and creative ideas about nature can be concocted. The bad thing is that the adaptationist paradigm does not provide for a means to rigidly test hypotheses about adaptation. An example is the notion (which has actually been advanced!) that hominids first stood up on their hind limbs to indulge in aggressive genital displays. The just-so story is its own justification; but, as we've seen, science isn't like that. That means that our notions have to be testable, which can make things difficult.

Often, traits are there for unforeseen reasons or exist as a result of some unrealized force or evolutionary event. And sometimes the trait that is under scrutiny isn't adaptive at all; it may even be detrimental to the organism, simply existing because something genetically associated with it is useful. An excellent example of this kind of phenomenon is the structure of the human skeleton, and in particular the structure of our knees and feet. The current structure is very useful to us when we want to run, and more than likely was selected for during human evolution. But it is a fragile structure prone to breaking and tearing easily; and, whatever else it may be, it isn't optimized.

So there is a tradeoff of benefit and detriment. If we want to understand our origins as humans, and how our uniquenesses came into existence, then better understanding and integration of modern evolutionary biology, genomics, and paleoanthropology is needed. As these fields draw closer over the next decades, this process will surely yield some amazing and enlightening results.

Evolution can also result in the random existence of traits in species (as we saw with drift in natural populations for the "guevodoces" trait). These traits arise in species not as part of a progressive process, but rather as the result of drift, or of association with something else useful. Such traits and the alleles responsible for them often lie dormant in the organism and in its genome for long periods of time, with no apparent evolutionary purpose or reason. Then bang! The trait turns out to be suited for some new purpose.

In this way, populations of organisms may be adapted in advance to particular environmental challenges. The traits and alleles that are "hiding" are called "exaptations," and many of the traits that we see as unique to modern humans seem to have arisen first as exaptations. A good example of a human exaptation is speech. The structures that allow speech most likely did not evolve as an adaptation for this specific purpose in early human populations. Instead they were in place well before we have any independent indication that they were used to project language. Speech thus capitalized post-hoc on the results of other forces, adaptive or otherwise, that were at work to alter the structure of the upper vocal tract.

Basically, evolution has to work with what's already there, and that means novelties that came into being randomly, via mutation. New uses for such novelties are not necessarily discovered right away, as we'll see over and over again when we recount the story of human evolution.

Evolution at Higher Levels

What goes on inside populations is critical to the evolutionary process, because it is only in this venue that the novelties arise that evolution can work on. But just as the sorting among genes within the population is undoubtedly important, so also is the winnowing of entire populations and species that takes place at the ecological level, for species don't exist and evolve in isolation. All species belong to communities of animals and plants within which there is a complex web of interactions.

One of the most important of these interactions is competition, and competition can often come from unexpected quarters. Environments and even geographies tend to be highly unstable, even sometimes over quite short time scales, and the actors in the ecological play are constantly changing. We're all aware of the havoc that has been wrought on faunas and floras around the world after the introduction, by people, of species that originated elsewhere. Snake-heads from east Asia are trashing what's left of the native fish populations in the southern United States, and the island of Guam has lost practically all of its native birds to introduced brown snakes. However well they got along before, native residents often have the greatest difficulty coping with new arrivals.

The examples we just gave are only a couple among thousands, if not millions; and even though human activities have resulted in enormous ecological changes in an unusually short space of time, climatic and habitat changes and animal migrations from one region to another have been a constant occurrence over the ages. In other words, the rules of the evolutionary game at this higher level are and have been constantly changing, and it's little use being the best adapted and most reproductively successful member of your kind if your entire species is being out-competed into extinction.

What everything we've just discussed about the genome and populations means is that when we look at the human fossil record to try to understand the background out of which we emerged, it's important *not* to see our evolution as a process of steady fine-tuning over the eons within a single lineage. This involves resisting the easy temptation to assume that just because we are the only hominid on Earth today, this is a normal state of affairs. It's even more important not to assume that, by extension, this means there has always only been one kind of hominid around. Instead, the fossil record tells us that the story of hominid evolution has been, from the very earliest days, a dynamic one: one of new hominid species coming onto the ecological stage and competing both with close relatives and other ecologically similar species for survival over the long term. Just as certain genetic variants disappear from the genome, an awful lot of species and local populations have gone extinct over hominid history.

Well, it's very easy to talk about this in the abstract. But for the working human biologist, issues such as these raise an alarming number of practical problems. For a start, if you are not simply following an evolutionary chain over time, and are instead trying to follow the complex plot of an evolutionary play with a huge cast of characters, you have to be able to identify the individual actors with some degree of confidence. As we've already hinted, this turns out to be far from easy. The reason is that, because species are biological populations consisting of individuals who vary from one another both genetically and physically, it regularly becomes necessary to decide when the differences you see between two fossils are simply those between individuals from the same species, or whether they indicate two different species—meaning essentially whether they belong to the same potentially interbreeding unit. It might seem straightforward to look at two closely related living species and use the degree to which they differ to the eye as a yardstick for judging this. But then you find that species are wildly variable in the amount of anatomical variety they enclose. Some species diversify physically to an amazing degree, while still maintaining reproductive cohesion. On the other hand, reproductive barriers can exist between species that look astonishingly similar. Add to this the fact that, as we've seen, at best, you generally have a pretty spotty sampling of species that are known only from the fossil record, and you can see how fraught with uncertainty the whole process is.

In paleoanthropology this basic business of species recognition is probably more contentious than in any other area of systematics, not least because the human fossil record has been much more intensely scrutinized than any other—and by scientists with a range of different evolutionary models in mind. As we've already mentioned, our view is that, under the continuing sway of the Modern Synthesis, paleoanthropologists probably tend nowadays to recognize fewer actors in the evolutionary play than there actually were. The range of reasonable perspectives is great enough that it is hardly realistic to expect consensus any time soon, if ever, making it all the more difficult to reconstruct the various acts of the human evolutionary play with as much confidence as we'd like. And it means that we often cannot even approach the task quite as rigorously as we would ideally desire, because to make testable hypotheses about the relationships among species you have to know with some confidence what those species were. Still, considerable progress has been made in recent years in refining the basic structure of the human fossil record; and, however approximate our categories may continue to be, new schema of human evolution are being proposed and tested all the time. Beyond this, the toolbox we described earlier has by now become so rich and resourceful that the scenario of human evolution we sketch later is not only a full and satisfying one, but has its feet firmly on scientific ground—despite its inevitable Panglossian (Calvinist) elements!

CHAPTER FIVE

THE PLACE OF *HOMO SAPIENS* IN THE TREE OF LIFE

A N UNDERSTANDING OF EVOLUTIONARY TREES AND HOW THEY ARE CONSTRUCTED IS ESSENTIAL TO AN INTERPRETATION OF THE PATTERNS OF DIVERGENCE

of our close relatives. By taking both a paleontological and a genomic approach to this subject, we can show with some confidence how humans fit into the hierarchy of life on this planet. Such an approach can explain the significance of similarities or "homologies" in structure from one organism to another. And it gives us a scientific avenue toward understanding the nature of the similarities and differences between ape and human features, such as the structure of the hands, the role of the vertebral column in upright bipedality, and the homologies between the various regions of the brain in humans and apes. An understanding of homology is critical to interpreting the common origin of the behavioral and anatomical attributes of organisms, such as communication, language, art and music. And to understand homology, we have to understand the Tree of Life and how it is constructed.

Systematics: The Tree of Life

Systematics is the field of biology that attempts to determine the relationships of living and extinct species in nature. Because all organisms on the planet have a single common ancestor and because evolution has occurred as a result of descent with modification from ancestors, the pattern of evolution can be depicted as a tree. As we've seen, the only figure in Darwin's *On the Origin of Species* is of an evolutionary tree. In fact, Darwin doodled in the margins of his notebooks frequently, and one of the most interesting of these doodles is an evolutionary diagram he sketched in 1842, showing the backbone of ancestral descendent relationships of organisms in the form of a branching diagram or a tree. [Figure 40] While Darwin wasn't the first to suggest trees as a way of depicting the relationships of organisms, his work stimulated many later scientists to use the analogy of a tree to represent ancestor-descendent relationships.

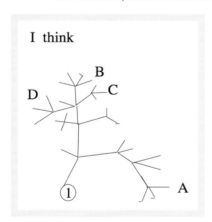

FIGURE 40. An electronic trace of the tree in Darwin's notebook. The "I think" above the tree actually appears in the notebook written in Darwin's own hand.

Today, an effort is being made to place all of the living named species in a tree-based context. The effort to construct a tree of ancestor-descendent relationships of all 1.7 million named species is called "Assembling the

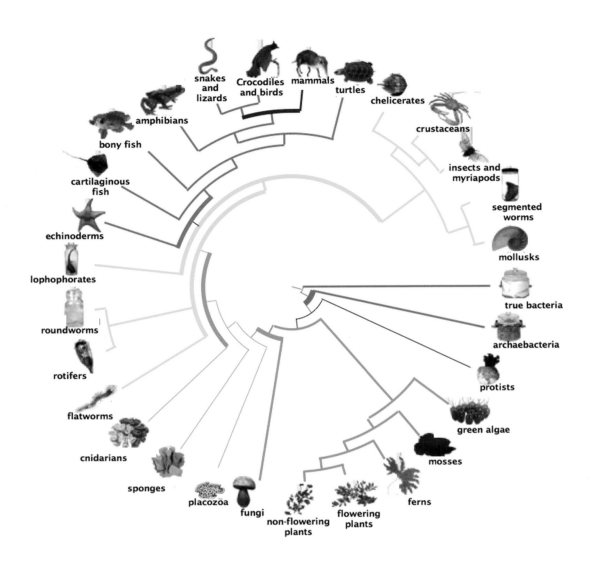

snakes
and
lizards

Crocodiles
and birds

mammals

turtles

chelicerates

crustaceans

insects and
myriapods

amphibians

bony fish

cartilaginous
fish

echinoderms

lophophorates

roundworms

rotifers

flatworms

cnidarians

sponges

placozoa

fungi

non-flowering
plants

flowering
plants

ferns

mosses

green algae

protists

archaebacteria

true bacteria

mollusks

segmented
worms

An Archea an Archea a
Kingdom for Archeae

Fungi Are Closer To Us
Than Just Between Our
Toes

Squish Squish

Anus First

Dem Bones

A Hairy Subject

FIGURE 41.

Red: An Archaea an Archaea, my kingdom for an Archaea. The base of the tree of life is one of the most interesting questions in biology. We have looked at the tree of life for a long time as having five kingdoms. Work accomplished by Carl Woese at the University of Illinois has shown that there are actually two kinds of prokaryotes—true bacteria and archaea indicated by the red branches in the tree.

Green: Fungi are closer to US than just between our toes! The green lines indicate the placement of plants and mosses and fungi. Note that the fungi and plants do not have a common ancestor in the circle tree. This means that fungi are actually more closely related to animals than they are to plants.

Orange: Squish Squish. The next major groups to diverge in the tree of life are the squishy things like sponges and jellyfish (cnidarians). These lineages are shown in orange.

Yellow: Anus first or mouth first? The next animals to diverge were animals with bilateral symmetry. What discriminates these animals from others involves their mode of development. In the animals on the yellow branches the mode of development is to develop their mouth first, while the rest of the animals, including us, develop anuses first.

Purple: Dem Bones, Dem Bones. The purple lineages indicate the divergence of vertebrates. Note that an invertebrate—starfishes and their close relatives—is part of this group.

Blue: A Hairy Subject. Mammals diverge from other vertebrates (blue branches). *Homo sapiens* is in this branch of the tree of life.

Tree of Life." [Figure 41] The nice thing about using a tree to represent the divergence of life is that the branch tips of a tree of this kind all have the same importance in the tree. The tips represent all living organisms, and sometimes extinct ones too. Representing them in this way places what we usually consider more simple organisms, such as bacteria, on the same evolutionary plane as what we consider more complex organisms, like ourselves. Contrast this view of life with the Scala Naturae view as promulgated by Aristotle, Linnaeus, and others, in which there is a rigid top-down ranking of living forms.

The Truth About Cats and Dogs

How does systematics work? The techniques of systematics have arisen through much trial and error over the past two and a half centuries. Prior to the 1960s, systematics was not rigorously scientific, as we have defined science earlier in this book. It was sort of accomplished by committee, and was called "evolutionary taxonomy." In this form of systematics, "experts" pronounced authoritatively on the relationships of these organisms or those, and if their reputations were great enough, their words on the subject were the last words. But such expert testimonies were, in reality, nothing more than untested and usually untestable hypotheses, even though based on observation and usually an acute understanding of the diversity of organisms.

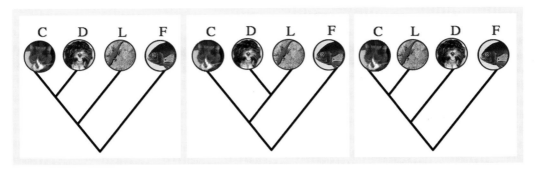

FIGURE 42. The truth about cats and dogs. The three ways a dog, cat and lizard can be arranged in a phylogenetic tree. Dog with Cat (left), Dog with Lizard (middle) and Cat with Lizard (right).

In the 1950s, some systematists started to question the expert approach and to develop methods that approached systematics from rather more testable perspectives. Then in the 1960s, in systematics as in almost every walk of life at that time, all hell broke loose. Whether it was the rebellious young upstart systematists of the 1960s questioning authority, or because systematics had reached an impasse, two major ways of circumventing the expert testamentary approach of evolutionary taxonomy began to be followed in the building of evolutionary trees. The first of these used similarity as a basis for classification. In this approach, organisms were classified together if they were more similar to each other than they were to other organisms. Thus, all of the observations made on the organisms under study were boiled down to some measure of similarity, which was then used to establish relationships. The second approach used the raw information from individual attributes of organisms. If a particular characteristic (or, more properly, state of a characteristic) was confined to a pair of species, for instance, it was taken as evidence of relatedness between them as a result of inheritance from a common ancestor.

The latter approach is the one that is nowadays preferred by most systematists, especially as it specifically allows for the extraction of characters that can diagnose groups of organisms. It is also based on a strong philosophical feature called parsimony. Parsimony is used in systematics to decide which of many possibilities is the best explanation for the data. For instance, if a particular explanation involves three steps in the change of a character, whereas another transformation takes five steps, then the former explanation is more parsimonious and is preferred. A decision about which phylogenetic tree is the best explanation for the data is typically based on choosing the one with the fewest character changes. This process by which we make evolutionary trees is called phylogenetics.

Now consider the following characteristics for a cat, a dog, and a salamander, where a 1 means the characteristic is present and a 0 means the characteristic is absent:

	Lungs	Hair	Legs	Mammary Gland
Cat	1	1	1	1
Dog	1	1	1	1
Salamander	1	0	1	0

Live trees need roots, and so do systematic trees. The root of a phylogenetic tree tells us which part of the tree is at the bottom (more ancestral) and which part of the tree is pointed up (more derived or unique). The root of any phylogenetic tree allows scientists to say whether a particular trait found in some of the organisms in the tree evolved in descendants in the tree or whether the trait was present all along in the tree. In this example, both hair and mammary glands are present in cats and dogs and absent in salamanders. But what is the ancestral state? Were these two traits present all along and just lost in the salamander? Or did the two traits evolve in the ancestor of cats and dogs? [Figure 42]

To "root" the tree for a salamander, a cat, and a dog, we can use an organism that we know is no more related to any one of the three species than to the others. A fish, which is equally related to all the others, is a good choice for this job. The following information on whether the trait is present (1) or absent (0) can be used for fish as our "root":

	Lungs	Hair	Legs	Mammary Gland
Fish	0	0	0	0

There are three possible ways to arrange a cat, a dog, and a salamander on an evolutionary tree with just these three species. We can put a dog with a salamander, or a cat with a salamander, or a dog with a cat. In this example, "lungs" and "legs" don't do us much good in determining which of the three possible trees is best explained by our characteristics, because dogs, cats, and salamanders all have lungs and legs. These two anatomical features tell us nothing about which species share unique characteristics. But "hair" and "mammary glands" are useful, because they are characteristics that are shared and unique to cats and dogs. If these four traits are the only four characteristics we can find that are evidence of relationships of cats, dogs, and salamanders, then the characteristics support the tree with dogs and cats together.

Now consider the following short gene sequences of dogs and cats and salamanders:

Cat	GGGTATATATATA
Dog	GGGTTTATATATA
Salamander	GGGTTTCTATTTA

And again, because the tree needs a root, we have the sequence for a fish:

Fish	GGGTATCTTATTA

Whether a particular column has a G, A, T, or C in it can now tell us which of the three possible trees is best supported by the DNA sequence. The two columns in red indicate positions that have DNA information that tell us that dogs and cats go together. In the first red position, cats and dogs are A, and other creatures are C. In the second column, dogs and cats are A, and all other creatures are T. The column in which the GATC's are underlined supports the relationship of dog with salamander. Dog and salamander have T's in that position, and the other creatures have an A in that position. Because there are two positions supporting dog plus cat and only one supporting dog with salamander, the arrangement of dog with cat wins.

With the advent of whole genome sequencing and high throughput methods for obtaining sequences (see Chapter 3), the amount of DNA sequence information available for use in systematics has grown exponentially. Twenty years ago, some very prescient scientists realized that DNA sequencing would produce a huge amount of information. These scientists established a national repository for DNA sequences called GenBank, a sort of bank for all of the sequences generated by scientists. Each sequence has a number (called an accession number), a name (the gene or genomic region name), and a description, and the actual DNA sequence of the gene or genomic region is given. GenBank is available on the World Wide Web and can be searched using a variety of vehicles. Figure 43 shows the number of gene sequences available in the national and international sequence databases. What large numbers of sequences in the database means for evolutionary tree building is that there are numerous tools for reconstructing phylogenetic relationships of organisms.

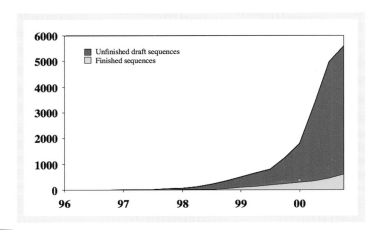

FIGURE 43. Number of sequences in Genbank for the Human Genome Project. Yellow indicates the number of raw sequences deposited into the database by year from 1996 up to the time the "first draft" of the genome was announced. Red indicates the number of sequences in the database of finished annotated sequences in the database by year. Redrawn from the GenBank website.

Now, imagine needing to construct a phylogenetic tree for four species and one root. In this case, you would need to evaluate 15 trees for the most parsimonious arrangement of these four species. For five species it would be 105 trees, for six species 995 trees and so on. By the time you hit a data set with 100 species in it, the number of trees you would need to evaluate is more than the number of particles in the universe. Obviously we need to have computers evaluate these problems. The trees that we settle on are not easy to interpret. Even if there are just a few species in each tree, we can be tricked by simply looking at the trees, because they have lots of branches and one tree might look like another when in actuality they are telling us completely different stories about the history of life (see examples in tree test below). We have already seen that evolutionary trees are similar to living trees, in that they have branches, trunks, and roots. But the similarity stops there. Whereas living trees have fixed branches, evolutionary trees have branches that can rotate. In order to test your ability to read and interpret evolutionary trees, we offer the following quiz based on one published in the journal *Science* in 2005. Once you have mastered the ability to read phylogenetic trees, we can then proceed to show you some of the more interesting and astounding results of phylogenetic analysis that have been worked out in the last decade.

Testing Your Tree Thinking

Cast of Characters

FIGURE 44. The cast of characters for the phylogenetic trees discussed here. From left to right – Fish (F=*Lobotes*), Cat (C=*Felis domsticus* [Mellow the Cat]), Dog (D=*Canis familiaris* [Miley the Dog]), Lizard (L=*Rhoptrophis bradfieldi* [Nambia Day Gecko], and primate (P=*Homo sapiens* [Erin the Gleavy]).

Answer each of the following five questions as best as you can. Score 20 points for each.

Question 1. Which of the following trees does not imply the same set of relationships as the others?

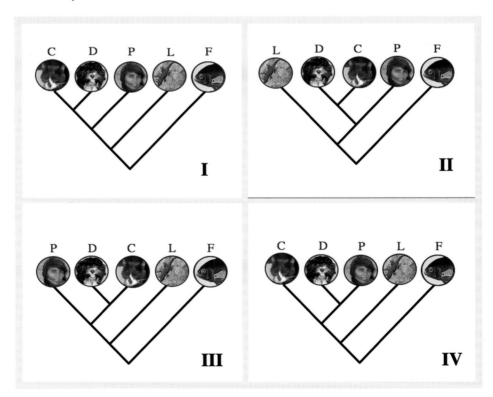

FIGURE 45. Question #1. Which of these trees is different from the others? I, II, III, IV, they are all the same, or they are all different?

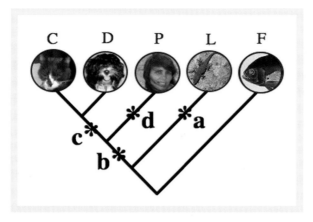

FIGURE 46. Question # 2. Where in this tree would a bonobo attach: a, b, c, d, or none of the above?

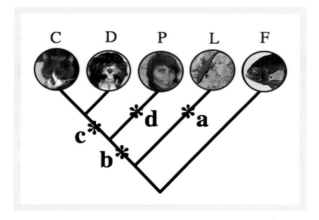

FIGURE 47. Question # 3. Where in the tree would a marsupial attach: a, b, c, d, or none of the above?

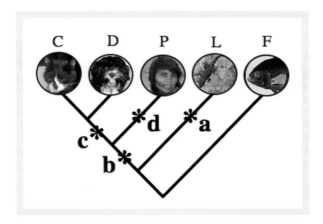

FIGURE 48. Question # 4. Where in the tree would a bear attach: a, b, c, d, or none of the above?

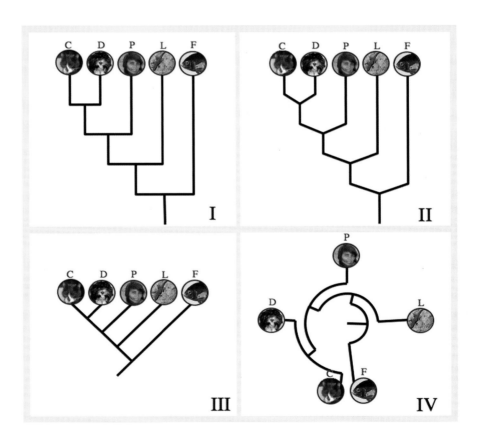

FIGURE 49. Question # 5. Which of the trees is different from the others: a, b, c, d, or none of the above?

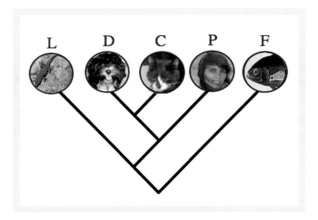

FIGURE 50. Extra Credit (add 20 points for a correct answer). A good root for the following species in the tree would be:

    a) an orangutan
    b) a tiger
    c) a snake
    d) a sea urchin

Answers. 1 (IV), 2 (d), 3 (b), 4 (none of the above), 5 (they are all the same), and Extra Credit (d).

If you scored 80 points or higher, then you are pretty good at thinking in trees. If not, then try tree thinking on other trees by going to the *Science* Web site: http://www.sciencemag.org/cgi/content/full/310/5750/979 (membership, subscription, or payment required). It really isn't that hard; it just takes a little practice.

We are now ready to clamber around the tree of life. There are many routes to take. The way we have chosen in this book is to bounce around the tree of life toward our own branch, in order to find out where in this convoluted, branching, crowded tree we sit. We call this exercise Clambering to Us.

An Archaea, an Archaea, My Kingdom for an Archaea

Before the use of molecular information in systematics, all organisms on the planet were thought to fall into several higher categories called kingdoms. One of the kingdoms was called Prokaryota and included all single-celled organisms without nuclei. The kingdom to which we humans belong was called the Eukaryota, and consists of all cellular life (single-celled and multicellular) with nuclei. In the 1970s, a microbiologist working in Illinois discovered that some of the prokaryotes were very different from others. In particular, Carl Woese, professor of Microbiology at the University of Illinois at Urbana-Champaign, found that the extremophiles ("lovers of the extreme"), a group of non-nucleated single-celled organisms that mostly live in extreme environments, belonged to a different group altogether from the classical bacteria. This group of single-celled organisms is now called the Archaea, to designate its members' very archaic status. The discovery of a species is something to crow about if you are a scientist, but to discover

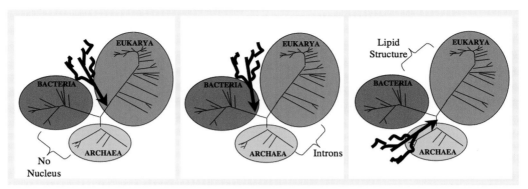

FIGURE 51. Where is the root of the tree of life? We see here the three pos- ROOT sible ways the eukaryotes, archaea and true bacteria could be related to each other. On the far left is the hypothesis that bacteria and archaea are each other's closest relatives, leaving eukaryotes out. This hypothesis is the classical prokaryote hypothesis, and fits well with the single celled non-nucleated nature of archaea and bacteria. In the middle is the hypothesis that archaea and eukaryotes are each other's closest relatives, with bacteria out. This hypothesis is supported by the fact that both archaea and eukaryotes have introns in their genes. The third and final hypothesis suggests that bacteria and eukaryotes are each other's closest relatives, with archaea out. This hypothesis is supported by the cell membrane structure of bacteria and eukaryotes. Scientists have settled on the middle hypothesis as the most likely of the three. Redrawn from www.palaeos.com/Kingdom/Images/Fig 01_16.JRG.

a brand-new kingdom of organisms is, simply put, amazing. Carl Woese is unique in this respect. [Figure 51]

There have been several ways to classify the major groups of organisms on the planet at the level of Archaea, Bacteria, and Eukarya. At one time they included five kingdoms, named Prokaryota, Protists, Plantae, Animalia, and Fungi. But any grouping of organisms we name should reflect a "real" group that includes all the descendants of a particular ancestor, and nothing but. The problem here is that Protists aren't a real group in this sense; and, as we shall see, Prokaryota isn't a real group either. Another way to ask if the prokaryotes form a real group is to ask if the Archaea are really archaic. In other words, are today's Archaea really the survivors of the group that gave rise to all others? It turns out that there are three ways to arrange Archaea, Bacteria, and Eukaryota, just like there are three ways to arrange a cat, a dog, and a salamander.

You can place Bacteria and Archaea together, just as the Kingdom Prokaryota requires. In this case, the major characteristic that supports this arrangement is the lack of a nucleus. But you can also place Bacteria with the Eukaryota, on the basis of their shared lipid bilayer structure. Finally, you can arrange the tree so that the Archaea and the Eukaryota are together; the character that would support this arrangement would be that the genes of both groups have an intron/exon structure (although not entirely identical intron/exon structures). Which of these is the best explanation of the branching order of life on Earth?

Well, to cut a long story (about 3.5 billion years' worth) short, it appears that Eukaryota (Us) are more closely related to Archaea than the Archaea are to the Bacteria. This arrangement of the major groups of living organisms suggests a less complicated hierarchy of life than the five-kingdom arrangement. Most students of the matter would thus conclude today that there are only three major groups, whether we call them kingdoms, or whether, as most prefer, they are known as domains. This arrangement of major groups is easier to visualize, and avoids the necessity of recognizing "unnatural" groups. For instance, if Archaea and Bacteria are placed different domains, then the necessity of naming Prokaryotes as if it were a real group is eliminated.

Fungi Are Closer to Us Than Just Between Our Toes

The Domain Eukaryota includes what used to be known as the Kingdoms of Animalia, Plantae, Protists, and Fungi. Eukaryotes are those organisms with membranes that surround their chromosomes, unlike bacteria and archaea, where there is no membrane surrounding the chromosome. Eukaryotes are unique in that they have specialized organelles in their cytoplasm. Animal cells have what are called mitochondria, small organelles that make energy in the cell. The mitochondria are believed to have been a particular type of bacteria (a proteobacteria) that was engulfed long ago by an ancestral eukaryotic cell. Plants have both mitochondria and chloroplasts (the things in plant cells that make them green). Chloroplasts are believed to have arisen as a result of an ancestral eukaryotic cell engulfing another kind of bacteria called a cyanobacteria. Chloroplasts, and, more importantly, as we will see later in this book, for our story of human relationships mitochondria, have the remnants of these bacterial genomes in them.

The classical way of looking at the relationships of these great groups of organisms is to say that the Protists are the most ancestral kinds of Eukaryota. Molecular work has shown that Animalia, Plantae, and Fungi are still real groups, because each has its own single common ancestor. It is now clear, however, that Protoctista is not a good natural group, because its members do not have a single common ancestor to the exclusion of all other groups. Some Protists, it turns out, are more closely related to the plant lineage, while others are more closely related to animals. But how are these groups related to one another? Fungi (mushrooms) and plants have classically been thought to be more closely related to each other than either is to animals, as witness the employment of scientists who study mushrooms in botany departments at universities, and in herbaria and botanical gardens. But using DNA sequence information, systematists have shown that Fungi is actually more closely related to Animalia than to Plantae. [Figure 52]

FIGURE 52. Clambering around the tree of life—Animals take form. Left, fungi; middle, mollusks; right, Arthropods. Fungus just doesn't grow between your toes, and, yes, those mushrooms in your salad are fungi.

Squish Squish: Animals Take Form

Animals are a diverse group and our home "kingdom." But the transition from single-celled life, to squishy life such as jellyfish and sponges, to higher animals, was a long journey that took about a billion years. What did the first animal look like? One way to visualize this first animal is to understand the phylogeny of all living and fossil animals; in other words, to use what we know about life and how it branched to reconstruct this ancestor. It is clear that there are several lineages of "squishy" things in several different higher categories that are called phyla. Basically, there are four major phyla of such animals—the Cnidarians (coral, jellyfish, and sea anemones), the Porifera (sponges), the Choanozoa (choanoflagellates), and the Placozoa.

Which of these four groups is the most ancestral? Molecular evidence points to the phyla of sponges and Placozoa as being the "mother" groups of all animals. The Placozoa are simple organisms with just three layers of cells, which is not too surprising, because almost all "higher" animals have tissues with three layers in them (though some animals have reverted to only two layers). The sponges superficially appear to be a single group but turn out to possibly consist of two great lineages that do not have a single common ancestor. In other words, sponginess may have evolved twice in the history of the tree of life. Next come the Cnidaria, the jellyfish, medusas and hydroids; and finally, the bizarre choanoflagellates fill out our squishy parade.

Then, something big happened. Not big in that animals grew larger, but rather in that the body plans of animals changed drastically. Animals began to develop into forms with what is called bilateral symmetry. For the first time, the body plans of animals look the same on both sides of a line drawn down their middle. What happened to cause this major change in how animals look? Molecular researchers have fingered a suite of genes called homeotic genes. These genes are found in all animals, but one special kind, called HOX genes, is responsible for axis formation in animals. The number of HOX genes expanded greatly in bilaterian animals, and scientists have shown that they take on the function of establishing axial formation of the body and limbs.

Anus First or Mouth First?

The next major step on the route we are taking to Us was the differentiation of the bilaterally symmetrical animals. During the differentiation of the Bilateria, early development could proceed in one of two ways. The first way of development was adopted by a large group of animals without backbones, called Protostomia. In these animals, the first opening of the embryo develops into its mouth. The group contains the invertebrates, such as arthropods and mollusks. This group multiplies its cells during development by a kind of cell division called spiral cleavage.

The other great group of Bilateria is called Deuterostomia. Their embryos develop their anuses first, followed by their mouths at the other end. Deuterostomes cleave their cells in a radial fashion during development. But while humans are deuterostomes, not all deuterostomes have vertebrae like us. One group of them, the echinoderms (starfish and sea urchins) have a completely different body plan from the vertebrates but form their early embryos in the same way we do. What this means, improbably enough, is that starfish are more closely related to us than they are to other invertebrates like insects, clams, lobsters, or crabs.

Dem Bones Dem Bones

This brings us to the vertebrates, the next great division we need to enter to clamber closer to Us. The most ancestral vertebrates around today are fish, but this does not mean that fish have evolved less than us or any of our closer relatives. In Figure 53, we see the distribution and relationships of several important groups, leading to the major group called Gnathostomes (clambering yet closer to Us).

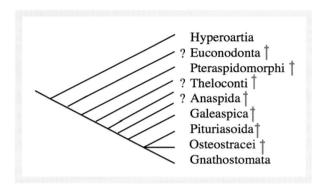

FIGURE 53. The divergence of the Gnathostoma .(things with jaws). The tree shows the various vertebrate groups that preceded vertebrates with jaws. They all have crosses after their names because none of these groups still exist. They are all extinct.

In drawing up phylogenetic trees, there is a convention that when a species name has a cross next to it, that species went extinct. Note that the great majority of major vertebrate groups at the level shown in the diagram are extinct, and also that all of these extinct vertebrates did not have jaws (Gnathostomata are the only vertebrates in this tree with jaws). Now, if we climb into the Gnathostomata part of the tree, we find ourselves alongside the fish and other forms that make up the jawed vertebrates.

In turn, there are two major kinds of jawed vertebrates—the Chondrichthyes (the sharks, rays, and other cartilaginous fish) and the Osteichthyes (our group), which can be divided yet again into two major groups—the Actinopterygii (the ray-finned fish) and the Sarcopterygii, our own group. The Sarcopterygii comprises the lobe-finned fish, plus the vertebrates that basically live on land. One of the most famous lobe-finned fish is the coelacanth. [Figure 54]

Our next major clamber in the tree of life is onto the branch containing the terrestrial vertebrates, which are a diverse group both in morphology and behavior. [Figure 55] Terrestrial vertebrates include amphibians, which do not have a watertight membrane

FIGURE 54. Clambering around the tree of life—Representatives of some major groups. Left, Starfish (Echinodermata); Middle upper, shark; Middle lower, Coelecanth, "living fossil"; Seahorse.

FIGURE 55. Clambering around the tree of life—More denizens. Clockwise from left upper. Sea tortoise; crocodile, snake, platypus, bird, tree frog, salamander. Of all of these the platypus is closest to US.

(amniotic sac) around their embryos. Typical examples of amphibians are frogs and salamanders. All other terrestrial vertebrates do have this sac and are called amniotes. This major groups of vertebrates includes turtles (Testudines) and lizards, crocodiles and birds, and mammals (Synapsida, including Us). By the way, if you are wondering where dinosaurs are (there are always one or two of you dino fanatics in the crowd), they are actually considered birds. Or rather birds are considered dinosaurs, and this places them both within the Diapsida!

A Hairy Subject

Mammals are the next big part of the tree of life we climb onto. The first branch in this part of the tree is Monotremata, a bizarre group of mammals that lay eggs. These strange animals lie at the very base of the mammal branch of the tree of life, and the best examples are the platypuses and the echidnas.

Next come the marsupials, a great group of mammals that lives mostly in Australasia. The really interesting fact about marsupials is that many of them, while living in total isolation from all other mammals, have converged on other mammals in shape, and sometimes in behavior too. For instance, there are marsupial "mice", "wolves", "moles", "bears" and "cats". The independent evolution of marsupials in Australia was one of the major pieces of information that led Darwin to reject the fixity of species. He realized that if two isolated and unrelated groups of organisms were showing similar structures and lifestyles, this must show that species can change through time to adapt to their environments.

Marsupials are the closest living relatives of the next group that we belong to—the Eutherian mammals, or mammals that have placentas. There are 18 living orders of placental mammals. The relationships of these orders has been the subject of intense systematic research, and several surprises have arisen during the past decade, as molecular and morphological information sets have been combined to give us a better-resolved picture of mammalian relationships.

A short diversion into whale systematics will demonstrate the power of combining molecular and morphological information. [Figure 56]

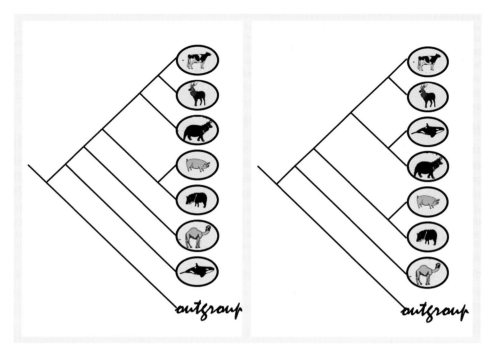

FIGURE 56. WHIPPOS!!. The phylogeny of mammals has benefited greatly from molecular information. One great example is the placement of cetaceans (whales) in the mammal tree of life. The tree on the left shows the relationship of cetaceans to the cloven-hoofed artiodactyls based on morphology. Note that the cetaceans are at the base or outside of all of the artiodactyls. The tree on the right shows the phylogeny obtained by examining DNA sequences of several genes. Note in this tree that WHales and hIPPOs (= WHIPPOS) are each other's closest relatives.

Traditionally, the closest relatives of the order Cetacea (the group containing dolphins and whales) has been considered on structural grounds to be the artiodactyls (animals like cows, pigs, goats, and hippos). But once we add molecular data to what we know about morphology, it appears that the mammal cladogram above shows us a more accurate representation of how whales are related to other mammals.

Look closely at this tree, and think about what it tells us. No, your eyes aren't playing tricks on you; the tree based on molecular data is telling us that whales and hippos are each other's closest relatives. And, even more surprisingly, whales are embedded deeply within the artiodactyls. We call this group, whales and hippos together, "whippos." Or at least we can refer to this group familiarly as whippos, because its real name, Cetartiodactyla (the fusion of Cetacea and Artiodactyla) is just plain too hard to say and write.

Once this grouping was accepted, and scientists began to look at the anatomy of whales and hippos a little more closely, the next development was very interesting, because the anatomical character that most strongly unites all artiodactyls is the bone structure of the ankle. And guess what? Whales don't have ankles, so they were excluded from artiodactyls based on the lack of the ankle structure …—until fossil whales were found with ankles! And guess what again? The ankle of the fossil whales has the precise

FIGURE 57. Clambering around the tree of life—Hairy subject. Echidna, elephant mouse, koala, bat (marsupial).

unique structure that scientists previously used to link all artiodactyls together as a "real" group. The fossil whales with artiodactyl ankles provide a link between living whales and artiodactyls.

Some of the more interesting relationships among mammals concern our place in the luxuriant mammal branch of the tree of life. [Figure 57] The orders of mammals that are most closely related to our order (Primates) in this tree are small scampering mammals and forms that fly or glide—the Scandentia (tree shrews), the Chiroptera (bats), and the Dermoptera (colugos or flying lemurs). Both anatomical and molecular data support a close relationship among the members of this group. But will the real closest relative to primates please stand up, or fly away? It seems that the tree shrews are the best bet for the closest nonprimate relative to Us.

Primates

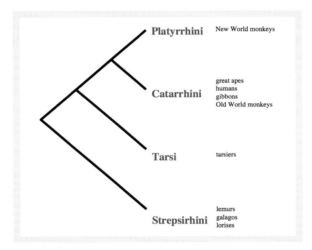

FIGURE 58. Monkeying around. There are four major groups of primates as shown by this tree. The Strepsirhini (lemurs, galagos, lorises) are at the base of the primate tree. Next comes the Tarsiiformes (tarsiers), with the New World monkeys diverging from a large group of primates including the Old World Monkeys (baboons, colobus, etc), and the gibbons, gorillas, chimpanzees, bonobos and humans.

We have been almost all over the tree, but still have a ways to clamber to get to Us—the primates. The primates are made up of about 250 species in four major groups. [Figure 58] Lemurs, lorises, and galagos are the primates most similar to our common ancestor, followed by the tarsiers. [Figure 59] Then, closer to us, are the New World monkeys like tamarins, squirrel monkeys, and capuchins. Closer to us yet come the wide variety of Old World monkeys and the apes. The Old World monkeys include primates like vervets, colobuses, and baboons. [Figure 60] And (drum

FIGURE 59. Clambering around the tree of life— Odd primates. From left to right; Tarsier, Lemur, Bushbaby, Potto.

FIGURE 60. Clambering around the tree of life—Monkey business. From left to right: howler monkey, vervet monkey, colobus, mandril, baboon.

roll, please), we finally come to the last five of our closest relatives in the tree of life—the gibbons and the great apes.[Figure 61]

It is very clear that gibbons were the first ape lineage to diverge after the Old World monkeys had gone their own evolutionary way. What is not entirely clear are the branching patterns of the last five great ape species—orangutans, gorillas, bonobos, chimpanzees, and Us. The current molecular data support the idea that orangutans were the first great ape lineage to peel off from the rest. What happened next is controversial, but a consensus from DNA sequence information has influenced how we think the events have unfolded.

Human-Chimp, Human-Gorilla, or Chimp-Gorilla

The very first molecular studies on humans and great apes did two important things. First, a branching pattern was suggested for the closest great ape relative of humans. The controversy involving these relationships revolved around determining which of the great apes—orangutan, chimpanzee, or gorilla (notice that our friend the bonobo is not included, simply because when these early studies were done, the bonobo was considered a subspecies of chimpanzee)—is our closest great ape relative. Several ideas based on anatomy existed prior to the onslaught of molecular data. The first suggested that chimpanzees and gorillas are each others' closest relatives and that humans are next most closely related to the gorilla-chimpanzee pair. A smaller number of scientists advocated the human-gorilla pair, or human-chimpanzee or human-orangutan, as closest relatives. The first molecular sequences resulted in a branching pattern suggesting that humans

FIGURE 61. Clambering around the tree of life—Our Closest Relatives. From left to right, top to bottom: Gibbon, Orangutan, Gorilla, Chimpanzee, Bonobo, Human.

and chimpanzees were each other's closest relative. The second important inference from early molecular work was the suggestion of a divergence time for humans from chimpanzees at about 5 million years ago.

The More (Sequences and Specimens) the Merrier

The inferences from the maternally inherited mitochondrial DNA molecule (we will hear about this molecule in detail when we talk about human migration) on branching order and divergence time were not radically new, but the mtDNA data were among the first generated from DNA sequencing to fully address what has been called the holy trichotomy problem (human-chimpanzee-gorilla). Since the first sequencing study using mtDNA, several other genes have been used to examine the problem. The chimp-human relationship is strongly supported when all of the newer molecular data from many, many more genes are examined alongside the earlier molecular data. The estimation of the divergence time has since been altered too, to reflect more accurate molecular clock estimates and newer fossil data. This estimate is 7 million years ago. But, oddly enough, while DNA sequence data support bonobos and chimpanzees (together) as our closest relatives, the morphological data from hard anatomical characteristics fail to corroborate this set of relationships clearly.

In the early months of 2006, while we were writing this book, a comparison of the chimp and human genomes suggested that there had been some traces of hybridization in the chimpanzee and human genomes. No, it's not what you think! This doesn't mean that humans and chimps have interbred. Rather, scientists who have examined the chimp and human genomes think they can see the traces of hybridization of the *ancestors* of chimpanzees and humans about 5 million to 6 million years ago. There are, however, a lot of problems with this inference, and more work needs to be accomplished before it will be possible to state with confidence that it occurred. On the basis of the fossil data, we prefer to leave the human vs. chimpanzee/bonobo divergence at 7 million years.

A Young Species?

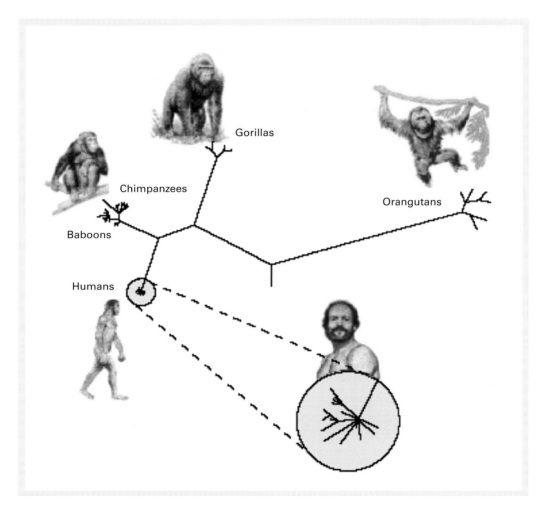

FIGURE 62. A young species. The tree shows the branching patterns of four of our closest relatives—orangutan, gorilla, bonobo, and chimp. The branch lengths are proportional to the amount of DNA change that has occurred since divergence from the common ancestors. Note that the branching within the *Homo* lineage is not discernable like the others in the figure – unless you magnify it. This pattern means that within *Homo* the divergence has been a very recent one relative to the other species in the tree. Courtesy of Srante Paäbo.

One problem with all of the early molecular studies of the closest primate relatives of *Homo sapiens* was the small size of the great ape and human samples used. With the advent of genomic high-throughput technology, many more samples from all of the great apes have now been examined for many genes. The Figure 62 tree shows the conclusions of one of the extensive studies of great ape-human relationships. In this tree two important things are going on. The first, and most obvious, is the branching pattern of the species in the tree. The second concerns the lengths of the branches, which are roughly scaled to time. A long branch indicates that a lot of time has passed since splitting from a common ancestor; a short branch indicates a shorter bit of time has passed since the split. This study also includes the bonobo, and it makes the holy trichotomy (Gorilla Human Chimp) problem a holy tetrachotomy—one, because the bonobo is now generally recognized as a species by itself. Perhaps the most important result of this study stems from looking at the lengths of the branches of the great ape tree. One group in the tree has very short branches emanating from its most recent common ancestor.

Guess which one? *Homo sapiens*! The reason for this is that our species is much younger than those grand great ape species. While the great apes may have fewer variants (fewer branches) than us humans, their branches are much longer, indicating a much older MRCA (Most Recent Common Ancestor). If we were to compete on who has been around longer, we modern humans would lose badly. Our most recent common ancestor is an infant compared to the chimp, bonobo, and gorilla's most recent common ancestors. How old is our lineage? To answer that, we have to go to the next big issue in human evolution: the "Out of Africa" question, discussed in the next chapter.

Homo troglodytes, Homo paniscus, and *Homo sapiens?*

Recently, scientists using molecular information have suggested that chimpanzees and bonobos are so closely related to humans that we should rethink how we have named the three species. The data that some scientists use to make this suggestion come from the comparison of the DNA sequences of the whole genomes of humans and chimpanzees, as well as from data on how genes are expressed in chimpanzees and humans. While the idea of placing chimps and bonobos in the same genus with Us is intriguing, we suggest that the classical generic divisions of *Pan* (for chimps and bonobos) and *Homo* (for Us and our close hominid extinct relatives) should remain as the best way to classify Us and chimps. Some of the reasons we reject the notion of recognizing Us and chimpanzees and bonobos as members of the same genus lie in the description of branching events leading to Us and chimps and bonobos since our common ancestor diverged. As we will see in the next two chapters, our lineage is rich enough with divergence events in and of themselves to warrant keeping it not only in its own genus, but its own family.

CHAPTER SIX

THE HUMAN EVOLUTION STORY

B Y ABOUT 7 MILLION YEARS AGO, THE CONTINENT OF AFRICA, FORMERLY CLOTHED BY FAIRLY CONTINUOUS FOREST, HAD BEEN EXPERIENCNG A DRYING

trend for some time. Rainfall became more seasonal, and woodlands with scattered trees, and even patches of grassland, were spreading as the dense forests began to break up. As the numbers of formerly diverse ape species dwindled, along with the habitat that had supported them, some of these hominoids (members of the primate superfamily that includes both the hominids and the apes) found themselves spending more time on the ground. It was out of this climatic stress that the hominid family was born.

In the period between 6 million or 7 million years ago and around 4 million years ago, several potential early hominid species are known, all of them discovered only within the last few years. Listing them is something of a baptism of fire for those averse to tongue-twisting zoological names, but unfortunately there is no way around using them, and for this we can only apologize. Four of the early hominid species (*Ardipithecus ramidus* and *Ardipithecus kadabba*, *Orrorin tugenensis*, and *Australopithecus anamensis*) are known from sites along the great African Rift Valley, while the fifth, *Sahelanthropus tchadensis*, is from the central-west African country of Chad. [Figure 63]

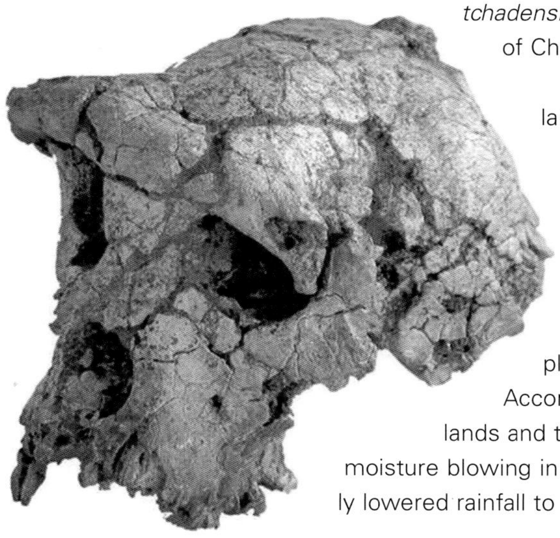

The recent discovery of the last of these laid to rest the "East Side Story" version of human origins which suggested that, as gigantic tectonic forces pushed the African landscape upward along the line of the African Rift Valley, from the Horn of Africa in the north to Mozambique in the south, the area to the east was placed in the rain-shadow of the highlands. According to the theory, the newly risen highlands and the land to their west trapped most of the moisture blowing in from the Atlantic, resulting in dramatically lowered rainfall to the east. To the west, the ancient tropical forest persisted; to the east, it fragmented into a mosaic of habitats in which the hominids emerged, while the hominoids of the western area remained forest-bound and gave rise to today's African apes.

FIGURE 63. The cranium (familiarly known as "Toumaï") of the putative early hominid species *Sahelanthropus tchadensis*, described from Chad, Central-West Africa, in 2002. Courtesy of Michel Brunet.

This makes a great story, but alas, as a scientific proposition it was falsified by the finding of the more than 6 million-year-old *Sahelanthropus* far to the west of the Rift Valley, in Chad. Clearly, the story of hominid origins was more complicated than the East Side notion made out, and it remains a story whose details are very hazy. Indeed, many hominids prior to the emergence of our own genus *Homo* around 2 million years ago are associated with faunas that suggest quite densely forested conditions.

Still, the one feature that all of the contenders for early-hominid status have in common is that each one was suggested by its describers to have been bipedal—to have walked upright on two legs. Ever since Darwin's time, suggestions have been bandied about as to what the adaptation was that truly marked the emergence of the hominids. At one time, for instance, the leading candidate was the big brain that is so characteristic of *Homo sapiens* today, but that was finally knocked on the head a half-century ago with the revelation that the big-brained Piltdown "fossil" was a fraud. Finally, the laurels have gone to upright bipedalism, which is, of course, the other really striking uniqueness of *Homo sapiens* among the primates; it is this that has become the *de facto* Rubicon that any fossil must cross to be regarded as hominid. Still, the early hominids that we have listed make a motley assortment, and in the case of all but the youngest of them, the 4 million-year-old *A. anamensis*, bipedalism has been established on fairly tenuous evidence. Indeed it seems likely that, with the thinning of the forests, several different hominoid lineages experimented with life on the ground, at least part-time.

Early Bipeds

Why a hominoid descending to the ground to exploit the resources made available by the new environment (and to compensate for the disappearance of the old ones) would choose to move upright on two legs is a much-debated question, for upright locomotion is a radical departure from the undoubted ancestral quadrupedalism. Many suggestions have been made as to the advantage the new kind of locomotion presented to the hominid ancestor. Some have suggested that it was more energy-efficient than moving on all fours, others that standing up freed the hands to carry things, or that it allowed predators to be spotted from farther away. Possibly the best current story invokes the high solar radiation that had to be coped with in the new, more open environments. By standing erect, this story goes, the early hominid reduced the surface area of its body directly exposed to the hot rays of the tropical sun, while at the same time the surface area available to dissipate body heat was increased. [Figure 64] This is a critical consideration, for overheating the body, and especially the brain, threatens survival.

Well, you can debate the pros and cons of standing upright forever, but the critical question to our minds is, why even try this bizarre posture? The most plausible answer is that the ancestral hominid already preferred to hold its body upright when moving around in the trees – as, indeed, several different primates do today. Chimpanzees are committed quadrupeds, and, like the other great apes, they have actually acquired a specific adaptation for moving on the ground on all fours, while retaining long, curved fingers that are useful for grasping branches in the trees. This involves curling up the hands and bearing the weight of the forequarters on the outside of their knuckles. But even so, some chimpanzees have been seen quite frequently to use upright locomotion when foraging from the ground in low branches of bushes and trees, as well as in the trees themselves.

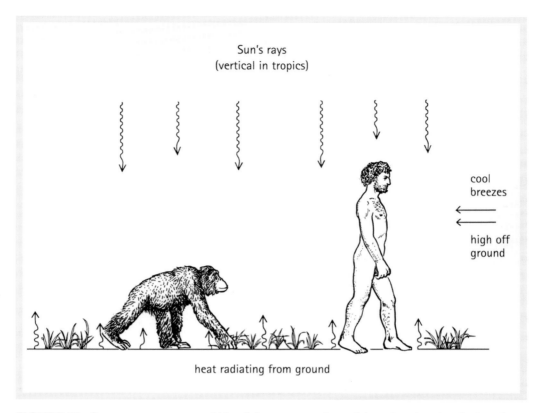

FIGURE 64. Some consequences of bipedal versus quadrupedal posture in a tropical environment. Compared to the quadrupedal ape, the upright human presents a smaller receiving area
to intense solar radiation and reflected and radiant heat from the ground. Cooling by evapora-
tion of sweat from the sheltered areas of the body is enhanced by wind effects high off the
ground. Illustration by Diana Salles.

So it's not at all unlikely that the ancestral hominid held itself upright when on the
ground—not an easy trick to pull off—simply because it already preferred holding its trunk
upright in the trees. It just felt most comfortable getting around that way, even in the new
milieu. And once it started moving on two legs while on the ground, it would have possessed *all* of the advantages—and disadvantages—that this new style of terrestrial locomotion conferred. This moves the question of advantage from the level of the individual
(Why did upright members of the population do better reproductively than the ones who
stuck to moving on all fours?) to the level of the species: Why did the species that chose
bipedality do better than the competition? Posed this way, the question awaits a satisfac-
tory answer. In any event, the several putative hominid species from this early time period show a pattern that seems to have held throughout the story of human evolution—right
up through the time of the emergence of *Homo sapiens*. This is not the pattern that was
predicted by the Modern Synthesis. What we see instead is a diversity of species, all experimenting with the evidently many ways of being a hominid.

As members of the lone hominid species in the world today, we take it for granted
that this is the "normal" situation. And this certainly fits well with the expectation of the
Modern Synthesis that the human evolutionary story was, largely at least, one of steady

improvement over time in a central hominid lineage. Yet it now appears that our lonely estate today is a highly unusual situation, and one that is probably telling us about the unprecedented nature of *Homo sapiens* rather than saying anything about the general hominid condition. And that, in turn, puts a pretty clear finish to our idea of ourselves as the perfected product of a long history of evolutionary burnishing.

"Bipedal Apes"

FIGURE 65. AL 822-1 from Hadar, in Ethiopia, is perhaps the most complete skull discovered so far of the early hominid species *Australopithecus afarensis*. Like its close relatives, and in complete contrast to *Homo sapiens*, it has a very large face protruding in front of a tiny braincase. Courtesy of Donald Johanson.

In the period after about 4 million years ago we have quite extensive fossil evidence of a variety of early hominids that flourished in eastern and southern Africa (with a western outlier, once again in Chad). These hominids are often known as "australopiths," after the genus *Australopithecus*, the first of them to be described, back in 1925, from a site in South Africa. The earliest species of *Australopithecus* is *A. anamensis*, which we've already mentioned. Known from fossils found in northern Kenya that are dated from between 4.2 and 3.9 million years ago, this is the earliest hominid that we know on the basis of good evidence (in this case, features of the knee and ankle joints) walked upright.

Still, the remains of this species discovered so far are pretty fragmentary, and the best-known example of its genus is the rather younger *A. afarensis*. This species is known from abundant fossils, dating from 3.7 million to 3 million years ago, that have been reported from several sites in Ethiopia and one in Tanzania. [Figure 65] Additionally, while the one 3.5 million-year-old fossil of *Australopithecus* known from Chad has been given its own species, *A. bahrelghazali*, many believe it should be included in *A. afarensis*.

Undoubtedly the best-known species of *A. afarensis* is the partial skeleton known as "Lucy" that was discovered by Don Johanson and colleagues in 1975. Even incomplete skeletons are vanishingly rare in the human fossil record, at least before the quite recent invention of deliberate burial, so Lucy gives us a very rare glimpse of the whole creature, or at least of a good bit of it. What this glimpse tells us—and it is amply confirmed by numerous less complete remains—is that although these early hominids were clearly bipedal when they were on the ground, they were not bipedal in quite the way we are today. For a start, they were rather small-bodied—Lucy herself stood not much over three feet tall, and although males were considerably larger than females, a four-footer would have been a tall individual. And their bodies were not proportioned quite like

FIGURE 66. A diorama at the American Museum of Natural History shows two early hominids making the 3.6 million year-old fossil trackways discovered in the 1970s at Laetoli, in Tanzania. Although clearly bipeds on the ground, early hominids retained many adaptations that would have served them well in the trees. Photograph by Dennis Finnin/AMNH.

ours: For example, their legs were rather short, and their shoulders were narrow, while their rib cages widened dramatically downward to match a very wide pelvis. This makes for an interesting combination of features, for while the narrow shoulders, relatively long arms, short legs, and long, slightly curving hands would have been very handy while moving around in the trees, the flaring pelvis is distinctly unlike that of the arboreal apes. Indeed, although the pelvis of quadrupedal apes is long and narrow while ours is broad, Lucy's pelvis is broader yet. The significance of this has been debated long and loud—indeed, it has even been argued that Lucy was "hyperadapted" to bipedalism; but a general consensus has emerged that what we are seeing here is a sort of "compromise" adaptation, one that allowed considerable agility in the trees while permitting upright walking on the ground.

And walk upright these creatures plainly did. It is rare indeed to have direct fossilized evidence of behavior, but this is precisely what is furnished by the extraordinary footprint trails discovered during the 1980s at Laetoli in Tanzania. [Figure 66] Three and a half million years ago, a pair of australopiths walked across a plain at Laetoli that was covered in a thin layer of volcanic ash that had been dampened by rain. Trudging across this sludge, which would have had the consistency of wet cement, these bipeds left long trails of footprints just as you or we might leave traces on a wet beach.

Soon thereafter, the prints were protected by a new ashfall that was only recently eroded away to expose them at or near the surface again. The exact details of the footprints are—inevitably—debated, and there is right now no consensus on exactly how "modern" the feet that left these markings were. But there is no doubt that the creatures who made them—very plausibly members of A. afarensis or something like it; fossils of this species were reported from not far away—were walking upright, in a straight line, confidently placing each foot right ahead of the other and not rocking from side to side as an upright-moving chimpanzee would. Laetoli, it must be said, is an unusual locality to yield evidence of an australopith, for the ancient environment around there was quite open, whereas most other A. afarensis fossils have been recovered with faunas suggesting a mosaic of riverine forest, woodland, and limited grassland. Most authorities would nowadays concur that the australopiths were generally committed to habitats with at least some trees, to which they were more or less tied for sleeping and for shelter from predators during the day, as well as for a large part of their sustenance.

Some later australopiths developed massive jaws and teeth that appear well-adapted for grinding roots and tubers that had to be dug from the ground (and some even seem to have used animal bones to help them in this pursuit); but fruit seems to have been a significant component of the diet of earlier australopiths. These were evidently pretty opportunistic creatures, who ate whatever was available; indeed, they have been described as "edge" species, exploiting the wide range of foodstuffs available in the interface zones between forest and grassland. This is a fairly hazardous ecological zone to occupy, and in ancient Africa it certainly teemed with predators of many kinds. The very first australopith to have been discovered, the infant from Taung, in South Africa, was probably the victim of a giant eagle. We have already mentioned the unfortunate South African australopith whose skull bears twin holes, almost certainly made by leopard canines.

Still, while they were undoubtedly upright bipeds, the australopiths were hardly what you would call "human" in other respects. Their brains, for example, were little better than ape-sized (around a third the size of ours, or even less), and those brains were housed in skulls that were distinctly ape-like in their proportions. As in modern apes, a very large and projecting face was hafted on directly in front of a small braincase – so small in some cases that, as among male gorillas, special ridges were developed on the outsides of the braincase to allow extra room for the attachment of large chewing muscles. This kind of conformation contrasts sharply with our own skulls, in which short, tiny faces are tucked below the front of huge balloon-like braincases. For this reason, many paleoanthropologists have taken to describing the australopiths as "bipedal apes." And certainly, nothing in the record would allow us to infer that in early stages, at least, they had developed a level of intelligence significantly greater than that of modern apes. All this said, though, it's important to note that these creatures were, in reality, neither humans nor apes. It would be a particular mistake to conclude that their physical structure or way of life were in any meaningful way "transitional" between the ways of doing business that are represented by modern apes and humans. Their way of life was something that was unprecedented at the time, and was evidently a highly stable and successful one.

In the long period between the first appearance of *Australopithecus* at around 4 million years ago and the disappearance of the last-known "robust" australopith species about 1.4 million years ago, numerous australopith species came and went, their remains documented in some abundance at sites from Ethiopia to South Africa. Yet their basic body structures and ways of living apparently did not change significantly. We have no evidence that, over time, australopiths edged significantly toward the physical constitution of later hominids, and there is no australopith that we can regard as anticipating human beings of modern body form.

The First Stone Toolmakers

FIGURE 67. Modern replica of an Oldowan core, in volcanic rock, shown surrounded by the flakes detached from it by blows from another stone. In most cases the sharp small flakes were the primary tools. Replica by Peter Jones.

We've mentioned that the archaeological record starts at some 2.5 million years ago, when the first stone tools start to show up at sites in Ethiopia and Kenya. These are fairly crude things, consisting of small sharp flakes of stone achieved by bashing one river cobble with another. But they were highly effective and, as we mentioned earlier, experimental archaeologists have butchered entire elephants using nothing more. Their invention must have made an enormous difference to the lives of the hominids who made them, allowing them to do many things that they had been unable to do before, such as to detach a limb from the carcass of a dead animal — always attractive to scavengers of many kinds, some of them pretty dangerous — and to retire to a safer place to consume it. [Figure 67]

Rocks were also used to bash limb bones, cracking them open to extract the protein-rich marrow within. But while the tools themselves are not too impressive to look at, making them was an altogether more remarkable achievement. For intelligent as modern apes are, experiments have shown that understanding the principles behind hitting one rock with another at precisely the angle that will give you a sharp flake instead of a useless lump seems to be beyond the capacity of any living ape, even with extensive coaching. What's more, the early toolmakers seem to have had a considerable degree of foresight. We know this because not all kinds of stone are equally good for tool making, and these hominids selected suitable stones and carried them around for quite long distances in anticipation of needing them in places where such stone types were not available.

So the invention of stone tools was clearly an epochal event in hominid history. Yet it appears that the toolmakers themselves were not physically any different from their non-toolmaking predecessors. At first, this might seem a bit counterintuitive, for wouldn't you need a bigger or better brain to make this intellectual leap? Well, actually, no. Any invention, even one as significant as this, has ultimately to be made by an individual, and to be somehow based on knowledge already there; and in appearance the individual concerned can hardly differ significantly from his or her own parents or offspring. Innovations always occur *within* species, simply because there is no place else for them to do so.

The example of the first stone tools announces a trend that we are going to see consistently throughout hominid evolution: Although the human fossil and archaeological records are both characterized by very sporadic innovations, new technological and new physical forms do not show up in sync. This seems to be true even though we don't actually know who the first stone toolmakers were. In the period between about 2.5 and 2 million years ago we know a variety of mostly pretty fragmentary fossils that have been assigned to "early *Homo*," but none of them actually shows evidence of any significant physical advance over the australopiths. Unfortunately, nowhere is there any definite association between early stone tools and the hominids who made them. The closest we come is at the 2.5 million-year-old site of Bouri, in Ethiopia, which has produced fossils assigned to the species *Australopithecus garhi*. Not far away were found animal bones bearing cut-marks that can only have been made by stone tools, and *A. garhi* is the obvious—but not demonstrable—culprit.

All in all, until we find some fossils from this early period that clearly belong to a more advanced hominid, we are more or less forced to conclude that the first stone toolmaker was a "bipedal ape" who discovered a behavioral potential that had previously been unexpressed, and thereby changed the course of history. No question, the invention of stone toolmaking did that, and the importance of this aspect of our behavior in demarcating us from the rest of Nature has actually been debated for centuries. Indeed, it is this preoccupation that explains why, in 1964, Louis Leakey and two colleagues assigned 1.8 million-year-old fossils from Olduvai Gorge to a new species *Homo habilis* ("handy man")—even though these fossils were hugely more ancient and more primitive than anything previously described as a member of the genus. Leakey subscribed to the ancient notion of "Man the Toolmaker," and he had explicitly been searching for the maker of the tools that he and Mary Leakey had been finding at the Gorge for decades.

But there's a problem with using unique behaviors to define species, and this is especially true when those behaviors turn out not to be unique, as Leakey's protégée Jane Goodall was shortly to discover, when she observed chimpanzees making tools from twigs to "fish" for termites in their mounds. Worse, though, the argument furnished by the morphology of the Olduvai fossils for including them in the genus *Homo* is actually pretty weak. Still, although Leakey had trouble getting his new species accepted, once his fellow paleoanthropologists finally took it on board, the floodgates were opened for a host of dubious inclusions in *Homo* that have confused the definition of this genus ever since.

Upright Striders

In 1891 a Dutch anatomist called Eugene Dubois found the skullcap of an archaic hominid on the Indonesian island of Java. It seemed to be pretty old, and its brain was quite large—twice australopith size, and well over half the size of ours today. But the braincase was long and low, and adorned in front by heavy brows. Because it was apparently associated with leg bones that looked pretty much like their counterparts in modern humans, Dubois called his find *Pithecanthropus erectus* (upright apeman). Today the specimen is called *Homo erectus*, and it has come to symbolize the phase of hominid evolution that lies between *Homo habilis* and the arrival of *Homo sapiens*. The original *Homo erectus* specimen is now believed to be about 700,000 years old, or maybe a little more, and fairly similar specimens from Java may span the astonishing range of about 1.6 million years to 40,000 years ago.

The fact that you have a single pretty recognizable hominid species in one peripheral spot in Southeast Asia for an incredibly long period might reasonably be taken to imply that what we are seeing here is the product of an evolutionary backwater. Nonetheless, the name *Homo erectus* has been liberally applied to hominids of very different morphology and age from throughout the Old World. The earliest of these come from Africa, where *Homo erectus* has been mostly reported from sites in the 1.9 million to 1 million-year range – significantly older than most from elsewhere. Actually, "African *Homo erectus*" fossils don't look awfully like those from Java, and although they make up a pretty motley assortment, it's useful to group them into the alternative species *Homo ergaster*. Most *Homo ergaster* fossils consist of jaws and teeth, and skulls in various stages of completeness that formerly held brains in the same general size range as the original Java skullcap. We thus know that they possessed faces and chewing teeth that, while still pretty large compared with our own, were substantially reduced in comparison to those of the australopiths. But we don't have many bones of the body skeleton – and when we do, we don't know which skulls they go with, with one dazzling exception.

In the mid-1980s, an extraordinary discovery was made in the sedimentary rocks to the west of Lake Turkana, in northern Kenya. Here, paleontologists discovered an amazingly complete skeleton of a *Homo ergaster* who had died by the shores of an earlier lake some 1.6 million years ago. Dubbed the "Turkana Boy" because the remains are those of a presumed male adolescent, this skeleton gives us a unique insight into the evolutionary state of hominids at this time. [Figure 68] Although in chronological terms this individual was only 8 years old when he died (developmentally he was rather older), he was already 5 feet 3 inches tall. And it's estimated he would have topped 6 feet had he survived to maturity. What's more, in addition to being tall he was slenderly built, and his limb proportions were much like those of tropical people today. Here at last, and pretty much unprecedentedly, was a creature whose body was built essentially as ours is. The principal importance of the Turkana Boy is to show us that, by 1.6 million years ago (and presumably earlier, though exactly how much earlier isn't certain), a hominid was on the scene who was finally emancipated from the woodlands and the forest edges to which his earlier relatives had been confined. The Turkana Boy's body proportions were ideal for shedding the tropical heat out in the shadeless savanna, opening up ecological possibilities for hominids that simply had not existed earlier.

While the Boy's body skeleton possessed most of the major innovations that characterize our own bodies today, from the neck up the story was rather different, as we've already suggested. Nothing about his archaeological associations suggests that this radical reorganization of the body was accompanied by any major cognitive change, despite a small increase in brain size. The Boy and his kin continued to make stone tools that were essentially of the kind their much more physically primitive predecessors had been making for almost a million years. And it is not until several hundred thousand years after

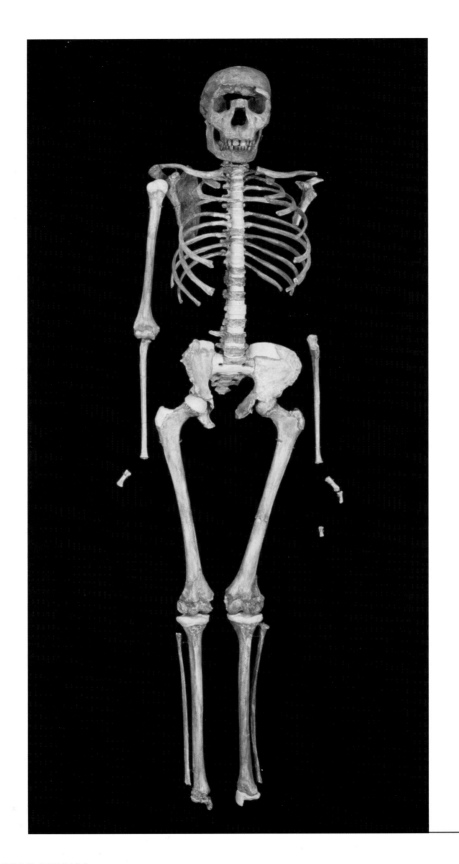

his species had first appeared that any significant innovation was made in stoneworking technology.

This innovation, when it came, was a big one. The early toolmakers were clearly not too concerned what the implements they made *looked* like. They were just interested in obtaining a sharp cutting edge. However, at about 1.5 million years ago (maybe a little more) hominids (again, we don't know just who, but in a broad sense it was surely *Homo ergaster*) began to make tools in a regular shape, most commonly that of a teardrop. The resulting symmetrical "hand axe" was formed using multiple blows to both sides of a piece of stone (later on, itself a large flake), and using a consistent technique. The shaping was evidently done to a "mental template" that existed in the mind of the maker before knapping began, and it clearly reflected another notching-up in the complexity with which hominids saw and manipulated the world around them. Unfortunately, important though this new development undoubtedly was, for want of evidence it's impossible to be sure how broad its impact was on the lives and the perceptions of the hominids involved.

Out of Africa

Hard on the heels of the point at which we can, for the moment, assume that the essentially modern body form had been achieved, hominids left Africa for the first time. Hominids are found at the site of Dmanisi, in the southern Caucasus between the Black and Caspian Seas, from some 1.8 million years ago. Most interestingly, it was not a remarkable increase in brain size, and thus presumptively of intelligence, that facilitated this migration into new and uncharted territories and environments. Several skulls are now known from Dmanisi, and all of them had brains smaller than the Turkana Boy's. [Figure 69] Indeed, one has a brain that is not much bigger than the largest found in the bipedal apes. Neither was it improved technology that permitted hominids to move beyond the tropics, because Dmanisi considerably predates the appearance of handaxes in Africa, and the stone tools found there are remarkably crude. This leaves only the new body form itself as a likely explanation of the newfound mobility of hominids, though we have to note that the few bones of the body skeleton so far found at Dmanisi, though as yet unpublished, indicate a much shorter stature than the Boy's.

Some very early dates have also been published for claimed hominid sites in southeastern Asia, though the records at these sites, or the datings themselves, are in most cases contested. But there can be no doubt that after their initial exodus from Africa (through population expansion, rather than through any form of deliberate expeditioneering), hominids in the form of *Homo ergaster* (as broadly conceived) and its descendants rapidly spread to far-flung parts of the Old World. Interestingly, and probably for reasons of physical and climatic obstacles, Europe was apparently occupied late: The earliest European hominid fossils, from sites in Spain and Italy, are only about 800,000 years old, and reliable archaeological indications do not go back much beyond 1 million years.

FIGURE 68. Skeleton of the "Turkana Boy." Dated to about 1.6 million years ago, this skeleton of an adolescent *Homo ergaster* from West Turkana in Kenya is the earliest we have that shows essentially modern body proportions. Photograph of mounted cast by Dennis Finnin and Craig Chesek/AMNH.

Still, the evolutionary pattern we perceive is very much what would have been expected for any new kind of

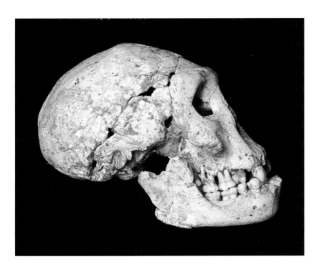

FIGURE 69. One of the five crania of *Homo georgicus* now known from the 1.8 million-year-old site of Dmanisi, in the Republic of Georgia. Though all are small-brained (less than 700 cc endocranial volume), they present a remarkable morphological variety. Courtesy of David Lordkipanidze.

mammal that was successfully exploring new territories far from its former heartland. New species were produced in different places: *Homo erectus* seems to have undertaken its own evolutionary as well as geographical exploration of Java and parts of China; and the newly discovered, tiny-brained (and disturbingly recent) *Homo floresiensis* from the Indonesian island of Flores is, if not an anomaly, an isolated island descendant of very early émigrés from Africa. Interestingly, before the recent discovery of handaxes about 1 million years old in the Bose Basin of China, these implements were barely known, if at all, from eastern Asia.

In Europe, early Spanish fossils have been described as the species *Homo antecessor* (for technical reasons, some think they should be called *Homo mauritanicus*), while an early cranium from Italy has been dubbed *Homo cepranensis*. This diversification is a totally unsurprising development that suggests that early hominids were playing the evolutionary game by an entirely routine set of rules.

Meanwhile, Back in Africa ...

Africa was not finished as a focus for paleoanthropologists once hominids had moved out of it for the first time. Far from it! It turns out that once the process had started, Africa served as an ongoing engine of innovation in human evolution, ultimately producing the hominid species—ours—that took over the world.

Back in the 1970s, the Ethiopian site of Bodo produced a skull of a hominid species known as *Homo heidelbergensis* that had long been known in Europe. [Figure 70] What was remarkable about the African specimen was its age: 600,000 years. The appearance of this species represented a huge stride in the direction of modern humans, the skull having generally a much more "modern" appearance than anything known from earlier. It was robustly built, but it had a less protruding face and better-inflated braincase than its predecessors did; and indeed, it had a pretty large brain, well within the size range of human brains today although still much below the modern average. Intriguingly, the skull concerned appears to have been defleshed after death by someone wielding stone tools, as witnessed by a series of cut-marks in the bone. Here we have a hint of the kind of murky behaviors we associate only with humans, and that are suggested earlier only by the Spanish *Homo antecessor* fossils from the Grand Dolina site. These were apparently treated in the same way as the animal bones found along with them, from which their discoverers concluded that their original possessors had been victims of cannibalism.

The Bodo skull was not accompanied by a very impressive toolkit. Indeed, the associated tools are surprisingly of the very early type, even though handaxes had already

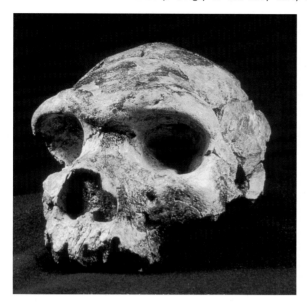

FIGURE 70. Cranium of *Homo heidelbergensis* from Bodo, in Ethiopia. Some 600,000 years old, this is the earliest known representative of a highly successful hominid species that appears to have spread throughout the Old World. Courtesy of Donald Johanson.

FIGURE 71. Reconstruction of one of the hut-like structures excavated at Terra Amata, in southern France. Perhaps 400 thousand years old, these structure were as much as 25 feet long. The cutaway reveals an interior containing a circular scooped-out hearth and the debris of stone working. Drawing by Diana Salles, after a concept by Henry de Lumley.

been used in its region some time earlier. So for more impressive technological achievements associated with *Homo heidelbergensis*, we have to look elsewhere. There are plenty of other places to look because, although it is first known in Africa, this species seems to have spread quite rapidly across the Old World: Representatives are known from as far apart as China and Europe. Fossils in Europe as old as 500,000 years belong to this species, and the best-known presumptively associated archaeological site is that of Terra Amata, on the Mediterranean coast of France. [Figure 71] Hunters 400,000 years ago seem to have established seasonal camps here, building the world's first well-documented artificial shelters. These were quite large structures, consisting of an oval arrangement of saplings embedded in the ground and brought together at the top to create a roof. The periphery of each hut was reinforced by a ring of stones, with a gap indicating where the entry was. Just inside the gap was a scooped-out area of floor, lined by blackened stones and containing burnt animal bones. This was a hearth, where a fire had burned under control and was evidently used for cooking meat.

The very first good evidence we have for the consistent domestication of fire in one place actually comes from Israel, where thick ash deposits have been found at a campsite some 790,000 years old. Then there is precious little evidence for hominid domestication of fire until Terra Amata times, after which the practice evidently became more common. The domestication of fire was a seminal event in hominid history, because the human

FIGURE 72. The finely-shaped point of one of the several 400,000 year-old wooden spears found preserved in a peat bog at Schoeningen, Germany. These carefully-shaped missiles have the proportions of a modern javelin. Photograph by C. S. Fuchs; © H. Thieme.

digestive system seems, on the face of it, not particularly well-suited to processing raw flesh, and neither are human teeth, for that matter.

This has led some to suggest that meat-eating really only got under way once fire had been domesticated for cooking, but we are now into murky territory, because chimpanzees occasionally eat meat, and it is almost certain that hominids had been consuming small vertebrates—lizards and so forth—for a very long time. Still, it reflects on the matter of hunting, which is a critical question because it has long been assumed that initial scavenging by bipedal apes (not necessarily of muscle meat—they may well have preferred organ meats, and we know that the very first stone toolmakers used stones to smash open long bones for the marrow inside) led ineluctably to hunting.

In this connection it is interesting to note that the 400,000 year-old site of Schoeningen, in Germany, has recently yielded some long, slender wooden spears that it is claimed were used as throwing weapons, because their weight is concentrated toward the front as in modern javelins. [Figure 72] This discovery was little short of miraculous, because wood is preserved for more than a few decades only under the rarest of conditions. The sophistication of these implements is perhaps greater than one would have expected from the associated stone tools, and reminds us that the stone tool record, splendid as it is, represents merely one aspect of hominid technological endeavor and gives us only the most indirect glimpse of the larger societies that produced it. Still, the spears were not tipped with stone, and would thus have had limited penetrating power when hurled. So the jury is still out on whether these were thrusting or throwing weapons. This is a pity, because if they truly were thrown, this would more or less confirm the arrival of sophisticated ambush-hunting of the kind carried out by historic hunter-gatherer peoples. As it is, we cannot be sure that this crucial innovation was introduced before *Homo sapiens* came on the scene.

However this may be, we have to wait until well after Schoeningen times to witness a true innovation in stoneworking techniques. This came with the development, probably in Africa, and after about 300,000 years ago, of the "prepared-core" tool. This kind of implement was made by carefully shaping a stone "core" until a single final blow would detach a virtually finished tool with a continuous cutting edge around almost all its periphery. We can assume, with limited confidence, that this kind of toolmaking was invented within *Homo heidelbergensis*, or at least by members of a very similar species; but probably the most impressive exponent of the method was *Homo neanderthalensis*. Named for the Neander Valley in Germany, where the first specimen was found, the Neanderthals were the product of an endemic hominid radiation in Europe; their bones have never been found outside Europe and contiguous western Asia.

We don't know whether either or both of the 800,000 year-old Italian and Spanish fossils signaled the beginning of continuous hominid occupation of the difficult European terrain, or whether they were representatives of a couple of failed early colonization attempts. But by about 500,000 years ago *Homo heidelbergensis* was established in Europe, and it may have persisted there for several hundred thousand years, though later dating is hazy. More certainly, by around 400,000 years ago hominids are found in Spain's Sima de los Huesos ("Pit of the Bones") that are distinctly different from *Homo heidelbergensis*, but that in some of their features clearly presage the Neanderthals. Whether or not these early members of the Neanderthal lineage were descended from an early population of *Homo heidelbergensis*, it seems clear that at 400,000 years ago there was more than one kind of hominid occupying Europe. *Homo neanderthalensis* itself only shows up at about 200,000 years ago, by which time it seems that this was the only kind of hominid left in Europe.

The Neanderthals

The Neanderthals are particularly interesting because, of all extinct hominids, they are the ones who have left us with the richest records of themselves and their lives. One major reason for this is that they lived widely in limestone regions where they often camped in the natural shelter of cave entrances and rock overhangs. In such places, the detritus of their lives (otherwise known as archaeological deposits) was often protected from erosion. Moreover, they also invented the practice of burying their dead, at least occasionally, and always simply. This is almost certainly why we have at least a handful of partial Neanderthal skeletons available for study, permitting a recent composite reconstruction of an entire Neanderthal skeleton. [Figure 73]

The Neanderthals were gifted toolmakers, making stone tools beautifully, although as one French prehistorian rather unkindly put it, "stupidly." He said this because over the huge expanse of time and space the Neanderthal inhabited, their tools were always basically the same – in dramatic contrast to the innovative habits of the *Homo sapiens* who eventually replaced them.

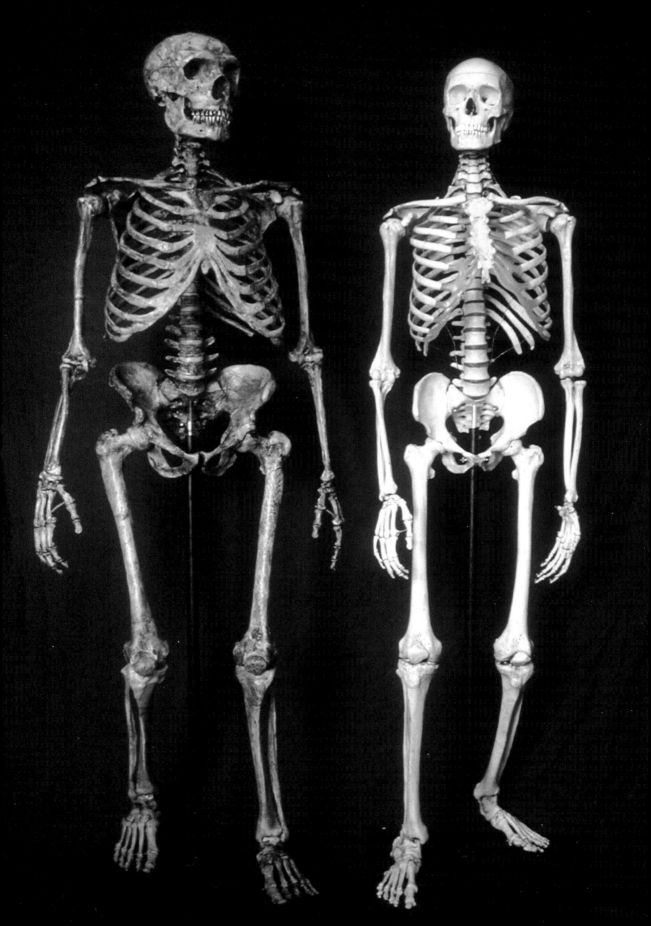

Recent studies, both of stable isotopes and of tooth wear, have suggested that the Neanderthals ate a lot of meat, which is hardly surprising given that they inhabited Europe during the Ice Ages when huge herds of grazing animals roamed the landscape, and plant resources were less diverse and abundant than they were farther south in the tropical and subtropical zones. The Neanderthals were obviously flexible and resilient, though; both from place to place and in the same place at different times, they adjusted their strategies to suit local circumstances, as we can tell from the animal remains they left behind.

But one thing we don't find in Neanderthal sites is much, if any, evidence of symbolic behaviors. This is particularly interesting because the Neanderthals had brains fully as large as ours, and our own behaviors are deeply symbolic. We create symbols in our minds and recombine them in our heads to remake our perceptions of the world around us, and the kinds of objects we make reflect that fact. In contrast, the record left by the Neanderthals looks pretty much like a more sophisticated version of those left by their predecessors. A handful of sites in a limited area of Europe show apparent modern human influence on Neanderthal material products after the arrival of modern humans in Europe and before the departure of the Neanderthals. But there is no reason to suppose that without outside influence the Neanderthals had developed "modern" ways of doing things–or, indeed, would ever do so.

Still, the Neanderthals are remarkable in what they achieved. They lived through some intensely difficult climatic ups and downs, and their technology was clearly up to the task of sustaining them under severe conditions. Physiological modeling has shown that there is no way they could have compensated biologically for the coldest climates they encountered—to achieve that they would have needed subcutaneous fat that weighed as much as the entire remainder of their bodies! Instead, the Neanderthals must have had substantial cultural resources to keep them going, of the kind that did not survive to form part of the archaeological record. Foremost among such resources must clearly have been warm clothing of some kind—most plausibly animal furs tied on with sinews – but the cold-weather kit would presumably have extended well beyond this.

The fact that the Neanderthals had coped so successfully in the past makes it unlikely that their disappearance some 30,000 years ago, shortly after modern humans arrived in their European homeland, and admittedly as Europe was in the grip of the last Ice Age, can reasonably be attributed to climatic stress alone. The only truly new feature of the Neanderthals' world when *Homo sapiens* arrived on the scene was *Homo sapiens* itself And almost certainly, the demise of the Neanderthals had much to do with the interaction between *Homo sapiens* and *Homo neanderthalensis*. Exactly what the nature of that interaction was is hard to say. Modern humans being as they are, it is hard to believe that direct conflict was not involved. But on the other hand, there are reasons for believing that the moderns were also superior economic competitors. The numbers of archaeological sites they left behind suggest that they existed in greater density on the landscape than the very sparsely distributed Neanderthals ever had; and this, by itself, suggests that the moderns were more efficient exploiters of natural resources. In the end, probably both factors took their toll.

FIGURE 73. Composite reconstructed skeleton of a Neanderthal, *H. neanderthalensis* (left), compared to a modern *H. sapiens* skeleton of similar stature. Note particularly the dramatic contrast in the proportions of the thorax and pelvis. Reconstruction by G. J. Sawyer and Blaine Maley; image by Ken Mowbray.

What we can be pretty certain of, however, is that the two kinds of hominid did not mingle, at least to the extent that would have led to any significant biological integration of the two populations. We can be fairly confident from the extent of the anatomical differences between them that the Neanderthals and moderns represented two distinct species, and thus would not have effectively interbred. The Neanderthal skull, though it contained a brain just as large as ours, was very different structurally. Neanderthals had large, protruding faces with curiously swept-back sides, and their braincases were long and low, receded behind large brow ridges, presenting very different profiles. Their body structure was different, too, notably in the proportions of the funnel-shaped rib cage and flaring pelvis and in the thick-walled limb bones with expansive joint surfaces. Such differences would have affected the Neanderthals' gait, which would probably have been stiffer than our own – another reason for thinking they and *Homo sapiens* would probably not have recognized each other as very desirable mating partners.

The Emergence of *Homo sapiens*

The people who entered Europe around 40,000 years ago and rapidly displaced the Neanderthals are colloquially known as Cro-Magnons, for the rock overhang in western France (the name means "Magnon's shelter" in the local patois) where their remains were discovered in 1868. They left behind a dazzling archeological record that shows without a doubt that these people were *us* in every functional sense of the world. We'll look at this evidence in some detail later, but for the moment the most important question to ask is: "Where did these new people come from?" For an answer we once again have to look toward Africa, for it is on that continent that we find the first intimations of both modern anatomy and modern behavior—though, significantly, not at the same time. And it is to Africa, too, that comparative molecular studies point us in the search for the place of origin of our species.

In Africa, there is a handful of fossils from the past couple of thousand years that look like *Homo sapiens* in most of their features, but not quite all of them. A broken-up cranium found in the 1960s at Kibish in the Omo Basin of southern Ethiopia, for example, looks for all the world like a modern human skull, with the exception that the areas above its eye sockets are not divided into two separate sections by a crease, as ours are, and its lower jaw does not bear a chin like ours. But if not *Homo sapiens*, it is pretty close. It has recently been redated to an astonishing 195,000 years ago by some fancy long-distance geological correlation. To paleontologists, this would have looked like an almost impossible antiquity, if it were not for the announcement not long ago of a 160,000-year-old skull at the site of Herto, farther north in Ethiopia. While it hasn't yet been properly described, this specimen seems an even better candidate for the status of an early *Homo sapiens*.

Other hominid specimens in the 200,000- to 100,000-year range, from sites farther south in Africa, have also been looked upon as early *Homo sapiens*, but most of them are either poorly dated or fragmentary, leaving lots of room for debate. At the same time, though, there are definite indications that in this time frame there were also hominid fossils in Africa that clearly were not *Homo sapiens*, so the long-standing pattern of several different hominids sharing the continent seems still seems to have applied.

Outside Africa, the earliest skeleton that can be unequivocally identified as *Homo sapiens* was found at a cave-entrance site called Jebel Qafzeh, in Israel—which, interestingly, has also yielded human fossils that look considerably less "modern." [Figure

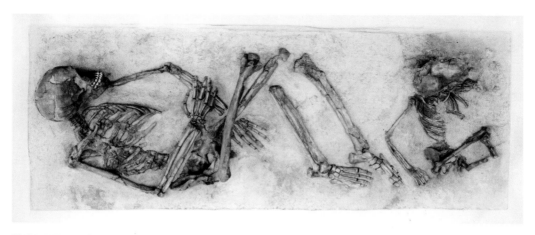

FIGURE 74. Over 90 thousand years ago a young female *Homo sapiens* was buried at Jebel Qafzeh, in Israel, with an infant, maybe hers, interred at her feet. Photograph courtesy of Bernard Vandermeersch.

74] The specimen concerned is, however, unquestionably *Homo sapiens*, bearing all the bony hallmarks of our species, and it has been thermoluminescence-dated as some 93,000 years old. Neanderthal specimens from the same region have been dated both earlier and much later than this, so in Israel there was evidently no pattern of abrupt replacement of one species by the other as there was in Europe. Rather, the pattern resembled what we see earlier in Africa, with several different early kinds of hominid around.

Very interestingly, all of these hominids were apparently behaving very similarly, at least to the extent that we can tell from the kinds of tools they made. Indeed, the very earliest claimants to be *Homo sapiens* made amazingly archaic types of tools: The Herto hominid, for example, is associated with handaxes, even though the Middle Stone Age, the African equivalent of the Neanderthals' "Mousterian" technology, had already been flourishing for quite a while. The few unremarkable flakes known from Kibish have also been attributed to the Middle Stone Age. At Jebel Qafzeh, again, the associated stone tools are Mousterian, pretty much indistinguishable from the tools that Neanderthals had made and continued to make in the region.

So in neither Africa nor Israel do we have any indication that the earliest *Homo sapiens* were behaving significantly differently from their predecessors. But at around 75,000 years ago, or perhaps even a bit earlier, we begin to find more "modern" behavioral expressions stirring. We have already mentioned that what marks modern *Homo sapiens* off cognitively from the rest of the world is its ability to think symbolically. The earliest intimations of symbolic activity have been reported from sites close to the southern tip of Africa. From Blombos Cave, Middle Stone Age archaeological layers around 75,000 years old have yielded ochre plaques engraved with a geometric design, as well as little snail shells with holes in them that have been interpreted as piercings for stringing—implying the making of body ornaments. There has been some argument over the interpretation of these objects, but their finders are satisfied that here they have evidence for some important aspects of modern behavior, even if in a rather primitive archaeological context. Somewhat more indirect evidence for symbolic behaviors has come from the rather older site of Klasies River Mouth, where archaeologists have evidence not only for cannibalism, in the form of burned and fractured human bones, but for functional divisions of the living space—something not previously a notable feature of hominid living sites.

Taken in conjunction with some other features of the record, such as evidence for long-distance trade in materials and the occasional making of long, thin "blade" tools rather like those made by the Cro-Magnons, there is an interesting case to be made for an early awakening in Africa of modern human consciousness, or at least of some aspects of it. It would, of course, be unrealistic to expect that the expression of this capacity of ours (as opposed to the underlying capacity itself) should have sprung into existence full-blown. After all, even today, we are still discovering new ways to deploy our intellectual potential.

As we've seen, the tendency throughout hominid evolution had been for several different hominids to share the world at any one time. It seems more than a coincidence that, in Israel, the Neanderthals continued to flourish up to the point, some 45,000 years ago, when a new technology appeared that was more or less equivalent to the one the Cro-Magnons took with them into Europe a few millennia later. At that juncture Neanderthals disappeared from Israel, just as they were about to do in Europe. And farther east, recent redatings make it appear that *Homo erectus* managed to persist in Java until about the time it is reasonable to suppose that behaviorally modern *Homo sapiens* arrived in the region. Even more remarkably, the diminutive and tiny-brained *Homo floresiensis*, the paleoanthropological shocker of the last decade, managed to hang on even later in its remote hideout on the Indonesian island of Flores.

The almost inevitable conclusion is that, in adding behavioral modernity to its already established modern physique, *Homo sapiens* had, in a very short space of time, become a uniquely dangerous competitor.

Again and Again; the Molecules Speak Up

The fossil data so far for the genus *Homo* clearly indicate that something interesting was happening among hominids in the period around 2 million years ago. The fossils also indicate a movement of modern individuals of the genus *Homo* out of Africa at least 1.8 million years ago. Recent molecular studies have shed much light on the scenarios we delved into in the early parts of this chapter, for in a later period we can ask the same big questions about our journey out of Africa, and attempt to answer them using molecules. Are modern humans the result of one or several moves out of Africa? If they are a result of more than one migration out of Africa, did the previous migrants mate with the new migrants? When did all of this happen?

The approach that has been taken is to obtain genomic information for as many *Homo sapiens* individuals from around the world as possible, and to compare the sequence information to construct genealogies and to infer coalescent events. But one of the more surprising results of the attempt to answer these questions using molecules is that scientists have been able to explore the mitochondrial genomes of several *Homo neanderthalensis* individuals. [Figure 75]

No! It's not what you might be thinking; these guys are *not* alive today. Scientists have accomplished some elegant and extremely careful experiments to obtain mtDNA genomic information from fossils. By drilling into Neanderthal fossil bone and then taking the powder produced by the drilling and isolating DNA from it, scientists were able to obtain enough Neanderthal DNA to get mitochondrial sequences. Unfortunately, it is difficult to isolate DNA from fossilized material, and many scientists have predicted that fossils that are much older than later *H. neanderthalensis* (30,000 to 40,000 years old) might not yield DNA. The reason for this is that DNA in dead tissues and bones degrades over time into tinier and tinier fragments, until what were once long chains of G's, A's, T's, and C's become single G's, A's, T's, and C's. So it has been suggested we probably won't get information for fossils much older than quite recent Neanderthals.

But then enter Mungo 3, a 60,000-year-old anatomically modern human from Australia. Mungo 3's mitochondrial DNA was analyzed, along with nine other (slightly younger) ancient human fossil remains from Australia. [Figure 76] The results of the analyses of Neanderthals and Mungo 3 are discussed in the next section and are somewhat controversial, but getting DNA from fossils or remains of people younger than Neanderthals or Mungo-3 is a possibility, and is beginning to be used to understand movements of people to all parts of Earth.

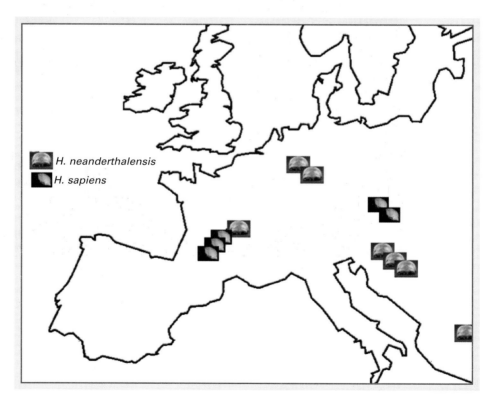

FIGURE 75. Localities of the Neanderthal and archaic human specimens that have been used in DNA studies.

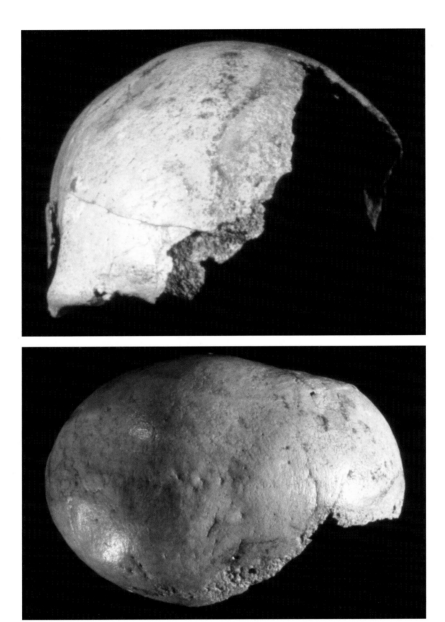

FIGURE 76. Front and top views of the Mungo 3 skull. Courtesy of Peter Brown.

What Neanderthal and Mungo 3 DNA Tell Us

Paleoanthropologists and geneticists can ask a slew of questions with respect to obtaining DNA from these important fossils. One of the hot-button questions about the movement of *Homo* species concerns whether our current species *Homo sapiens* is the result of a single migratory event out of Africa, or whether *Homo ergaster* (or whatever) came out of Africa long ago and migrated to the various regions of the world (Asia, Europe,

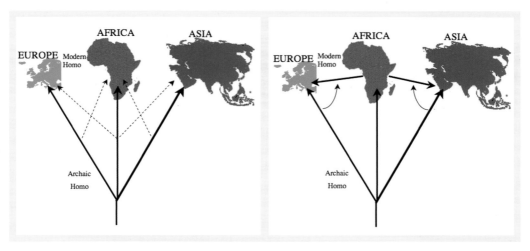

FIGURE 77. The multiregional hypothesis (right) versus the replacement hypothesis (left) of human colonization of Earth. The arrows show time getting younger from bottom to top. In the replacement hypothesis, earlier *Homo* who had first migrated to Europe and Asia are replaced by younger *Homo sapiens* migrating from Africa (solid lines) about 40,000 years ago. In the multiregional hypothesis, earlier *Homo* and younger *Homo sapiens* populations followed in single trajectories the three continental regions. No replacement occurred in this hypothesis and the three regions evolved independently of each other, although gene exchange (dotted lines) between regions kept the species unified.

and Indonesia), after which archaic humans in the four major regions of the world evolved into modern folk while at the same time maintaining local peculiarities due to geographic distances. [Figure 77]

The former notion is called the replacement hypothesis, and the latter the multiregional hypothesis. The viability of the multiregional hypothesis depends on whetheror not there is a pattern of shared genomes among all archaic humans. If the genome of, say, *Homo neanderthalensis* turns out to be more closely related to some *Homo sapiens* than the various *Homo sapiens* are to each other, then the replacement hypothesis is rejected, and the multiregional hypothesis would be more viable. If, however, *Homo neanderthalensis* genomes appear to be less related to those of *Homo sapiens* than the various *Homo sapiens* are to each other, then the multiregional hypothesis would be rejected and the replacement hypothesis more viable. Or does it have to be as simple as one or the other?

To date (2006), eight *Homo neanderthalensis* individuals have been sequenced for mtDNA. In addition, many early *Homo sapiens* have been sequenced. And then there is Mungo 3. By comparing these fossil *Homo* sequences with more than 3,000 living *Homo sapiens* sequences from all over Earth, scientists are able to tell the following story: The modern or living human mtDNA sequences show that all *Homo sapiens* make up one large group that excludes all eight *Homo neanderthalensis*. Mungo 3 has a DNA signature that makes it *Homo sapiens*, but it is a unique genomic signature that is found nowhere else in either other archaic humans or in living humans. Mungo 3 also has a very different molecular signature from the other nine ancient Australian *Homo sapiens* studied. These data suggest that *Homo sapiens* more than likely did not interbreed with Neanderthals; and that Mungo 3, a *Homo sapiens* individual, represents a lost genetic lineage that was living in Australia before the current Australian Aborigines arrived.

Laurent Excoffier, a biologist at the Computational and Molecular Population Genetics Laboratory at the Zoological Institute of the University of Bern who uses theoretical approaches to infer evolutionary events, has also taken data of this kind and made some very interesting predictions about Neanderthal-modern human interactions. Excoffier first looked closely at the mtDNA genealogy of living humans. This genealogy, as we will see shortly, traces maternal lineages of humans. He also examined the genealogy obtained using Y chromosome sequences, which follow paternal or male lineages.

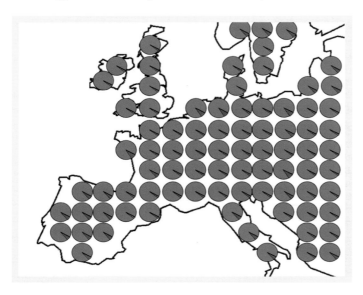

In both trees, African genes are found uniquely at the base of the tree (and not so uniquely elsewhere in the tree). His results indicate that all human maternal and paternal lineages originated in Africa and subsequently moved to other regions of Earth. While these data seem to support the replacement hypothesis, the ages of the lineages and more detailed analysis of more than just paternal and maternal lineages were needed to make a strong statement about replacement.

FIGURE 78. Population genetics simulation showing the percent of Neanderthal genes (red) of the total that would have persisted in European populations if Neanderthals had freely interbred with *Homo sapiens* in areas of contact.

Using population genetic approaches, Excoffier suggested that one could go back in time and predict how *Homo sapiens* might have interacted with *Homo neanderthalensis* when they came into contact. [Figure 78] Did they interbreed? If so, how much? He was able to suggest that Neanderthals and *Homo sapiens* did not interbreed. Two important inferences can be made by jumping into Excoffier's time machine. First, the pattern of replacement over generations can be simulated using the genetic information. In this simulation we can see that the *Homo sapiens* lineage takes over from and replaces the Neanderthal lineage in under 400 generations (about 10,000 or so years). *Homo sapiens* seem to have displaced Neanderthals and colonized Europe and the Near East like a bad disease in a science fiction movie. It is obvious that Excoffier's results support the replacement hypothesis.

Another interesting aspect of this theoretical work involved asking how much admixture of Neanderthals and *Homo sapiens* would be needed to leave some imprint of Neanderthal genes in current-day humans. In this analysis, Excoffier looked at different potential levels of interbreeding of Neanderthals and *Homo sapiens*, and found that even very rare interbreeding events would leave a Neanderthal footprint in the current genomic pools in modern *Homo sapiens*. Because we don't find this Neanderthal

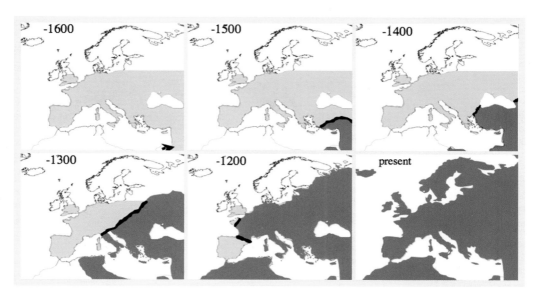

FIGURE 79. Population genetics simulations showing the spread of *Homo sapiens* (in purple) over Europe replacing Neanderthal (light blue) populations. The heavy dark lines indicate zones of "contact" between Neanderthals and sapiens. The numbers in the upper left hand of the six boxes indicate generations in the past.

genomic footprint, we can assume that Neanderthals did not interbreed with *Homo sapiens* as our species replaced them. [Figure 79]

Biologically, the next major phenomenon in human prehistory was the expansion of members of a formerly sparsely scattered species to our current vast distribution across Earth. This story starts with how we migrated out of Africa (again and again), moves to a chronicle of the 60,000 years (or 22 million days) that it took for modern humans to populate the globe, and ends with our current status as the dominant species on the planet.

CHAPTER SEVEN

LUCK AND HARD WORK

HOW DO WE TRACE OUR ORIGINS USING MOLECULES? IS IT JUST A MATTER OF LOOKING FOR SIMILARITIES AND DIFFERENCES THAT TELL US WHEN

and where we started to diverge from other primates, or when and where humankind started to move across Earth?

It turns out not to be that simple. But fortunately, nature has equipped our genomes with some unique features that help with discovering our origins through our molecules. Some of them took hard work to discover and analyze. Others were there for the picking. For instance, nature has left us with two tools in our genomes that allow us to trace female and male lineages: the mitochondrial genome and the Y chromosome, respectively. Scientists were lucky to have these two excellent and perfectly suited aspects of our genomes for evolutionary studies to focus on. Of course, genes on other chromosomes in our genome also have the potential to tell us a lot about our origins, but these take much more careful interpretation. One of the more important of these other kinds of inherited molecules, and fortunately relatively easy to understand, is found in the X chromosome.

Of Y's and When's

Each of your cells has a nucleus and a complete complement of 22 autosomal chromosomes (chromosomes not involved with sex determination) and, for most of the readers of this book, two sex chromosomes. Females have two X chromosomes. These X chromosomes are two large DNA molecules that are about 155 million nucleotides in length. Males are different, having one X chromosome and another, smaller, one called a Y chromosome. The Y chromosome is only about 50 million nucleotides in length. There are over a thousand genes on the X chromosome, while the Y chromosome is very gene-poor. But, while the Y is puny compared to the X (about one–fifth the size), it is still important for male sexual determination and development, or at least most males would say it is important, because this drives home the idea that size doesn't matter. But while most guys don't realize it, their Y chromosome is slowly deteriorating.

The Y chromosome of mammals was originally perfectly normal with respect to chromosomal lifestyle. In fact, about 100 million years ago in the common ancestor of all mammals, the Y chromosome was most likely a lot like our current X chromosome. But then it happened. It was perhaps a mutation or a change in the regulation of a gene on the Y-chromosome, and its effect was to change how sex determination is controlled in mammals. Whatever it was that happened, changes occurred in the switches involved in determining the difference between male and female fetal development. Because of this, the Y chromosome started to lose the ability to communicate with other chromosomes. Which is just a fancy way to say that the Y chromosome lost the ability to pair up, or recombine, with other chromosomes in the genome. This loss of recombination eliminated its ability to counter deleterious mutations that occurred in its genes, for recombination is an efficient way to cross-correct mutations that are deleterious to the fitness of an organism.

As the lack of recombination intensified, the genes on the Y chromosome either went missing from the genome or moved to other chromosomes, so that what was once a home for thousands of genes is currently reduced in humans to only 27 genes. We can easily trace the deterioration of this chromosome across evolutionary time by looking at the number of genes on the Y chromosomes of other mammals.

As time goes on this is rotten news for the Y chromosome, because if this loss continues, the Y chromosome will eventually be eliminated from the human genome. Brian Sykes of the University of Oxford and one of the gurus of human genetics, estimates that, based on the rate of Y chromosome decay, a 99 percent decrease in current fertility will happen in about 5,000 generations, or 125,000 years. Sykes predicts that the way the human genome determines maleness and femaleness will either change so that male determination is on another chromosome; or, worse for men, male determination will be eliminated from the human genome so that there will be no more males. In this strange scenario, humans will not be able to reproduce sexually, because there will only be one sex – female. Still, right now we retain the Y chromosome, and it has proven to be a powerful tool in understanding the peregrinations of human populations around Earth. In the following sections, it is the first tool we will describe for understanding ancient population movements because, as it is inherited from father to son, it is perfectly suited to tracing male lineages.

The Mightychondria

The second molecule we use in human origins studies is the mitochondrial genome. Outside the nucleus of most of our cells, but inside its walls in the cytoplasm, are tiny complex structures where energy for the cell is produced. These minuscule factories are called mitochondria; and while they are interesting structures in and of themselves, their origin in the cytoplasm of eukaryotic cells is even more fascinating. Apparently, as we've seen, mitochondria arose as a result of the incorporation (by eating, engulfing, or some other mechanism) of bacteria into early eukaryotic cells billions of years ago. Because the bacteria incorporated into the eukaryotic cells had a genome, mitochondria have retained some of that original bacterial genome. The mitochondrion and its genome have coevolved with eukaryotic cells for billions of years. This coevolution is so important that mitochondrial genomes have recently been shown to be very important in human health, as several human genetic disorders originate as a result of lesions in the mitochondrial genome.

Each and every mitochondrion in your cells has a small circular genome of about 16,000 base pairs, with about 13 genes coding for proteins. Each and every genome in these mitochondria—and there are thousands of them per cell—is, for the most part, a clone of the others; that is, they have identical sequences. This means that although there are millions of mitochondria in your body, there is only one mitochondrial genome and, more importantly for evolution, only one kind of mitochondrial genome in each animal sperm and egg cell.

This clonal inheritance of mitochondrial DNA is bad news for guys again, and here it is. Since sperm carry very few mitochondria, and these don't get into an egg when fertilization occurs, men do not pass on their mitochondrial DNA. In other words, men are dead ends for mitochondrial DNA. But because mitochondrial DNA is passed on from mother to daughter, it has become a perfect tool for examining the maternal lineages in human origins.

The X-Files

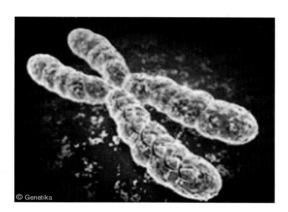

FIGURE 80. Electron micrographic image of an X chromosome.

We will also use the X chromosome to examine human relationships. [Figure 80] Every human being on the planet has an X chromosome. For the most part, biological females have two of them and biological males have only one. The X chromosome is one of the larger chromosomes in our genomes. It is over 150 million nucleotides long and has 1,098 genes on it. An unusually large proportion of genes on the X chromosome are prone to manifest genetic disorders.

Why? Because of the unique pseudo-pairing of the X chromosome with the Y chromosome in males. For instance, no matter whether you are male or female, your mother's chromosome 19 pairs with your father's chromosome 19 in most of the cells in your body. But if you are a male, in the majority of these cells your mother's X chromosome pairs with your father's Y chromosome. In females, there are fewer problems, because one of your mother's X chromosomes pairs with the only X chromosome your father has. So in males, there is only one chance to get an X chromosome gene right. What this means, guys (as if losing your Y chromosome wasn't bad enough), is that men are "X-challenged." So what? you might ask. Well, it means that males are more prone to expression of deleterious traits caused by defective genes on the X chromosome.

The Population Genetics of Genomes

With 25,000 or so genes in the genome, what makes X chromosomes, Y chromosomes and mitochondrial chromosomes so important to our story of human evolution? The answer is that all three of these entities are found in smaller numbers in the gene pool than other chromosomes that produce genomes in the next generation, and all three have interesting patterns of transmission from parent to offspring that make them perfect candidates for following human evolution.

Half of a human population will be males who each will have an X and a Y chromosome. Males will make sperm with either an X or a Y in them. The normal transmission of autosomes is much fairer, actually completely fair, in that there is no skew in the number of any of the 22 autosomes as they are distributed in sperm and eggs. Let's count

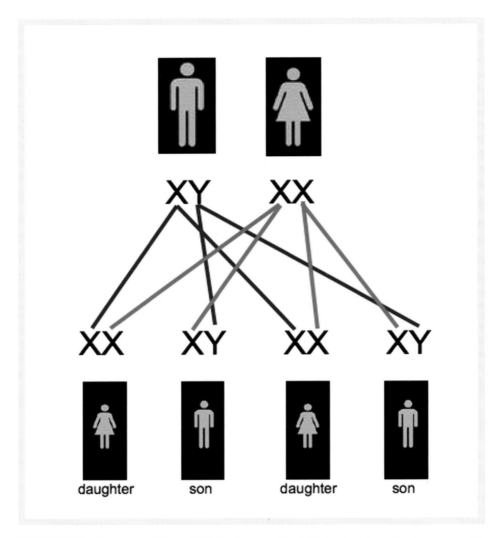

FIGURE 81. Diagram of X and Y inheritance. Most biological females have two X's and most biological males have an X and Y chromosome. Note that females get one X from their biological father and one X from their biological mother. Males on the other hand, get an X from their biological mothers and a Y from their biological fathers.

things up. Let's say a female makes two eggs and a male makes two sperm. The female will make eggs with X chromosomes in them. The male will make one sperm with an X and one sperm with a Y. The count here is X = 3 and Y = 1. [Figure 81]

For autosomes, the count is straightforward, four total of the same autosome in the two sperm and the two eggs. The X chromosome will therefore on average be found in smaller numbers in the reproductive cells of the next generation than the 22 autosomes – 25 percent fewer X's than autosomes make it to the gene pool. For the Y chromosome, the story is even more extreme. Because females don't have Y chromosomes to pass on to their offspring, their eggs will always have X's in them and, as we saw above, even for males, only half of their sperm will have Y's in them (the other half will have X's in them). So let's count things up again for two eggs and two sperm. A female will make both eggs with no Y chromosomes in them, and a male will make one sperm with an X and one with a Y. That's X=3 and Y=1 again.

The Y chromosome will be found in far fewer reproductive cells in the gene pool than the autosomes, making its population size far smaller than that of the autosomes. The smaller population size of both the Y and the X chromosomes means that these two chromosomes might be more susceptible to drift effects than the autosomes in the genome. Remember that genetic drift is that evolutionary force that results in smaller-sized populations accruing mutations that are not necessarily advantageous.

The mitochondrial genome is a different story. With respect to the number of mitochondrial genomes that are passed on from generation to generation, first remember that there is only one mitochondrial genome per egg and sperm cell. Couple this with the fact that the sperm cell does not contribute mitochondria to new offspring, and the population size of mitochondrial genomes contributed to the next generation after mating is, like the Y chromosome, only one fourth that of the autosomes in the cell.

Why do males not contribute mitochondrial genomes to the next generation? Because in the sperm the mitochondria are situated pretty far away from the head of the sperm that makes contact with the egg cell during fertilization. The egg cell also does a pretty good job of destroying any foreign mitochondria (i.e. those from the sperm) during fertilization. So egg cells maintain a strictly mother-to-daughter transmission of mitochondrial genomes. In addition, because of the smaller population sizes of mitochondrial genomes relative to autosomes during transmission, bottleneck effects will also be important in studying maternal transmission of this small genome.

The Pedigree of Woman (and Man)

In human genetics the study of pedigrees is extremely important in understanding diseases. Pedigrees represent the patterns of reproduction in families and help to explain how mitochondrial DNA and Y chromosomes can give us genealogies of humans. When you draw a pedigree chart, males are represented by squares and females by circles. Lines drawn horizontally between a square and circle represent a mating, and lines drawn vertically from the matings indicate offspring. Each generation's offspring is listed as male (squares) and female (circles), on the same level of the pedigree.

Just as medical scientists use pedigrees to examine disease transmission, so evolutionary biologists use pedigrees of genes (called gene genealogies) to examine the genealogy of lineages. For the most part, two kinds of lineages have been examined in human evolutionary biology: mitochondrial DNA to follow maternal or female lineages, and Y chromosomes to follow paternal or male lineages. These two tools of molecular evolutionary biologists have been used for the past two decades to understand human maternal and paternal divergence and to trace how *Homo sapiens* has spread across Earth, or rather, we should say, to understand how female *Homo sapiens* has spread by looking at mitochondrial DNA, and how male *Homo sapiens* has spread by looking at Y chromosomal DNA.

Figure 82 is a pedigree of one of the authors, going back to his great-grandparents and ending with his offspring (who, by the way, are adopted and female). In this case, his Y chromosome wouldn't have been passed on to his daughters anyway, so no big deal.

And his mother's mitochondrial DNA cannot be passed on by him anyway, so again this doesn't matter.

We have drawn the last generation with small circles and squares so as to get all of the offspring on the figure. It is very easy to trace the great-grandfather's Y chromosome throughout the pedigree. On the great-grandmother's side, there is no tracing to be done because the Y chromosome over there is going to be her husband's and doesn't make it through the maternal part of the diagram. Here is how the great-grandfather's Y chromosome gets traced. It is pretty easy to do because all you need to do is to follow the squares.

On the other side of the pedigree, we can trace the mitochondrial DNA of the author's great-grandmother. [Figure 83] This is done just as easily as tracing the Y chromosome, but it doesn't mean as much because, even though his mitochondrial DNA comes from his great-grandmother, and even if his children *were* his biological children, neither of his daughters would have received his great-grandmother's mitochondrial DNA. Fortunately for this author's mitochondrial DNA, he has two sisters who both have female offspring, and he also has several female cousins who have transmitted his great-grandmother's mitochondrial DNA. So the mitochondrial genome of his great-grandmother is not lost. But the trace is interesting, and also simple to make, as you just need to follow the circles:

Figure 84 shows the mitichondrial and Y chromosomal trees side by side. Notice that they tell different stories, based on the same pedigree. The very same principles of viewing Y chromosomes and mitochondrial DNA are in action when scientists use these tools to examine how females and males moved across Earth, only on a much larger scale. The big difference is that we don't know the exact evolutionary pedigree, as medical scientists do for their studies of human disease. The medical scientists who use pedigrees get

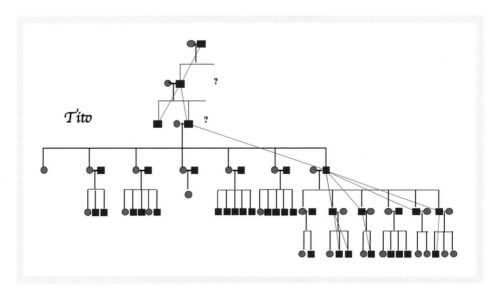

FIGURE 82. Pedigree of one of the authors on his father's father's side. Males are indicated by blue squares and females by red circles. Marriage of a male with a female is represented by a short horizontal line between a red ball and blue square. Offspring from marriages are shown as coming from the horizontal lines. The blue line traces the fate of the Y chromosome in the pedigree.

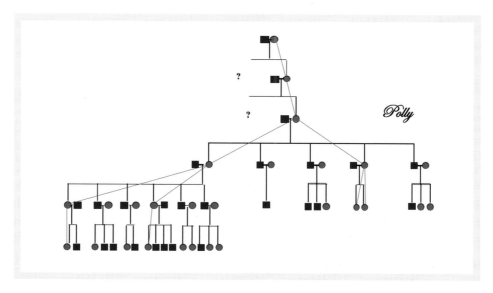

FIGURE 83. Pedigree of the same author on his mother's mother's side. The red line traces the fate of the mitochondrial DNA in the pedigree.

theirs by asking their patients who their parents and grandparents and their offspring are. For mtDNA and Y chromosomal DNA, the pedigree is inferred using the DNA sequence changes, in exactly the same way the tree of life is constructed. The only difference is the following: Remember that the tree of life represents the patterns of hierarchical divergence of species through evolutionary time. Mitochondrial and Y chromosomal DNA trees do not represent the genealogy of species, or even of people, but rather of lineages. Another way to look at it is to recognize that Y chromosomal DNA represents only a little more than 1 percent of the genome (50 million base pairs). You can then ask how it could possibly tell us how entire populations of humans have diverged. You can ask the same for the mitochondrial genome, which is only 16,000 base pairs long and represents only 0.001 percent of the genome.

So how can we use mtDNA and Y chromosomes to trace male and female lineages? Because mtDNA and Y chromosomes don't recombine! There is nothing for mtDNA to pair with in the cytoplasm; and, as we have seen, the Y and X chromosome do not pair properly. The rest of the 22 chromosomes in our genomes do, so if we use them as a source of data to try to tell how people have diverged, we find that they have recombined so much that the resulting diagram would be a net or a tangled mess—albeit a beautiful tangled mess!

Eve and Adam? Rebecca and Luca!

During the early days of using molecular markers to examine human evolution, origins and movement, mitochondrial DNA was used pretty much exclusively, because it was both easy to isolate from cells and easy to sequence. Many of the early studies contained landmark discoveries of the relatedness of human maternal lineages. The first, and perhaps most important, was done by Rebecca Cann and her mentor Allan Wilson at the

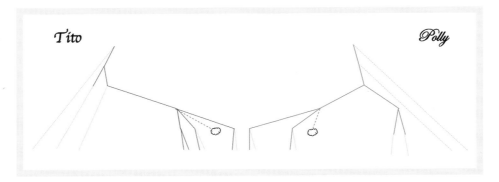

FIGURE 84. Full traces of the fates of the Y chromosome and mitochondrial DNA of one of the authors. The cloud puffs indicate the "extinction" of the marker in the author's lineage. Because the author in the pedigree represented here has two daughters, his Y chromosome is not transferred to his offspring (daughters are XX). The puff for mtDNA occurs because only females transmit mtDNA to the next generation. Fortunately for the author he has several brothers who have sons and two sisters with daughters so close versions of the author's mtDNA and Y chromosome will be transmitted to the next generation. Note also that the shapes of the two trees are different. This pattern is exactly what is seen when Y chromosome and mtDNA genealogies are examined over long evolutionary periods.

University of California at Berkeley. They demonstrated that, based on mitochondrial DNA sequences, the maternal lineages of a large number of humans indicated a single common ancestor for the human female lineage. They even suggested that the common ancestor had lived about 200,000 years ago. The common ancestor was unfortunately dubbed "Eve" and caused some furor in the press. But, as with all hierarchical lineages, a single common ancestor is a necessity of the analysis. The real situation was not that a single female existed that propagated the human lineage, but rather that a single maternal lineage *now* exists that can be traced back to the population that resulted in ancestor of all *Homo sapiens*. This does not mean that there was a single human female 200,000 years ago who propagated all humans we now see on the planet. The term "Eve" has since been used to designate major mitochondrial DNA lineages of people alive today. To date, there are about 18 "mitochondrial Eve" lineages that are also called "haplogroups" and are designated with letters A through M and T through X. These letters were applied in alphabetic order to the newly discovered human mitochondrial DNA haplogroups as they were found.

The Y chromosome was developed as a marker for males sometime after the mitochondrial DNA, because it was harder for scientists to manipulate and analyze. But by 2001, with the sequencing of the entire human genome, the tools to examine Y chromosomal DNA had become easy to use. A little after the "Eve" lineages had begun to be characterized, "Adam" lineages were being characterized in detail.

Peter A. Underhill and Peter J. Oefner, in Luca Cavalli-Sforza's lab at Stanford University, and Mike Hammer's lab at the University of Arizona, were the Y chromosome pioneers. Cavalli-Sforza, in particular, had been involved for decades in using genetics to understand human movement across Earth, and decided in the 1990s that the Y chromosome would be an excellent marker to examine for tracing male lineages. The work of his and other groups has deciphered the movement of males across the globe, using the approaches we have described for mitochondrial DNA. Not surprisingly, Adam lineages appear to have followed Eve lineages as humans populated Earth. The only difference is that only 10 Adam lineages have so far been detected in human male populations. Shortly, we will delve into the patterns of human Y chromosomal and mtDNA divergence and movement in some detail.

Clocking Evolution

While the first goal of using molecular markers is to decipher relationships and patterns of divergence, the second is to try to place divergence events in time. Paleontologists can use the regular radioactive decay clocks of certain elements found in the rocks surrounding fossils, and occasionally in the fossils themselves, to help calibrate divergence times. Paleontologists can also use the layers in which the fossils themselves are found to help give an age to certain divergence events. Molecular evolutionists use two rather controversial approaches to figuring out the ages of events—molecular clocks and coalescence analysis.

The first—molecular clocks—is an old method that was first suggested by molecular biologists who thought molecular sequences behaved like isotopes and were accumulating changes in a clock-like fashion. This idea was refined and championed by Allan Wilson in the 1970s. Molecular clocks from different molecules or genes were said to accrue mutations or "ticks" at different rates. For instance, genes that make proteins such as histone (a protein that helps keep eukaryotic chromosomes together) that are extremely important to all eukaryotic cells, and are thus highly conserved at the sequence level, will tick very slowly. These slowly changing genes will have perhaps one tick every 20 million years. Other genes that are not so important to all eukaryotic cells, such as genes involved in blood function like a protein called fibrinogen, might tick much faster with, on average, a change every 0.01 million (10,000) years.

Scientists use sequences from the different genes for answering different evolutionary questions. The histone gene would be like a grandfather clock we might use to tell the time of day accurately. Scientists use the histone clock to answer questions about the timing of the divergence of classes of vertebrates from each other (all vertebrates diverged from each other in the last 400 million to 500 million years). The blood protein fibrinogen would be like a stopwatch used to time a 400-yard dash. Scientists use fibrinogen for answering questions about the timing of the divergence of orders of mammals (all mammals diverged from a common ancestor in the last 100 million years).

So for human evolution, most of which occurred in the last few million years, an extremely fast-ticking molecular clock or set of clocks needs to be found; in line with our clock analogy, we need to find a highly fine-tuned stopwatch. Fast-ticking molecular clocks exist in both the mitochondrial DNA genome and in the genes on the Y chromosome. Mitochondrial DNA has a region called the D Loop. The D stands for "displacement" and denotes the part of the mtDNA genome where the genome starts to replicate. Because it doesn't need to make a protein, this region is not as highly constrained to stay the same as other genes in the mitochondrial genome, so it accumulates a lot more changes than any other region of the mitochondrial genome.

As for the Y chromosome, we know that eukaryotic genes have introns and exons in them. The exons are regions of the gene that are made into protein and the introns are regions within the gene that are not made into protein and that are snipped out in the nucleus before being translated to protein (see Chapter 3). Because they aren't there to code for the protein, the introns can accumulate changes much faster than the exon regions can. In this case, the introns in genes on the Y chromosome can be used as molecular clocks to calibrate divergence events along male lineages.

Molecular clocks are controversial, so their results are almost always interpreted with caution. Part of the controversy stems from the fact that molecular clocks are very irregular in their "ticking," which is very unlike the highly uniform radioactive decay of, for example, radioactive carbon 14. For most DNA sequences the irregularities tend to balance out over long periods of time, and fairly good estimates of divergence times can be obtained for more remote events. But one thing for sure is that if you have a fossil date and a molecular estimate for a divergence, take the fossil date. Why? As we just noted, the ticking of a radioisotopic "clock" is more regular and accurate than the ticking of molecules.

As a result, other methods have been developed to estimate divergence times. These methods have the funny but apt-sounding name of coalescence methods. Coalescence simply means "the union of diverse things into one body or form or group."[Figure 85] In the case of the divergence of species and populations, coalescence is viewed in retrospect. Instead of things coalescing *together* through time, we trace divergence events *backward* in time until they coalesce. Where they coalesce is called the most recent common ancestor, or the MRCA. It turns out that both population size and mutation rate have a huge effect on the time to coalescence, so the process of determining the time to the MRCA of lineages involves estimating what are called demographic parameters. In addition to population sizes and mutation rates, generation times (the time it takes to be born and then to reproduce) and other demographic parameters are also used to calculate coalescence times. Because these coalescence parameters can be estimated, calculation of the time to the most recent common ancestor is approachable without recourse to molecular clocks and the vagaries that plague molecular clock estimates.

Now that we have acquired the understanding to use the three different kinds of molecules (mitochondrial DNA, Y chromosomal DNA, and X chromosomal DNA) to look at human evolution and to calculate divergence to most recent common ancestors, let's take a trip around the world in 60,000 years.

Again and Again, Again

In 2002 Alan Templeton performed a rigorous analysis of several genes from many humans to re-examine the Out-of-Africa hypothesis. His analysis was meant to clarify the events or processes that had occurred to produce the current patterns of human genomic variation. In the process, Templeton also had a lot to say about the multi-regional vs. replacement question, and about placing dates on the major events involved in the evolution and establishment of the current human populations across the planet. His study used the coalescence methods we described earlier. Using 1.7 million years ago as the date that humans left Africa, and a coalescence analysis of several genes, Templeton suggested that at around 500,000 years ago another wave of migration occurred, of another species in the genus *Homo*. This migration was restricted, but the analyses suggest that these *Homo* individuals interbred with the descendents of the original wave of *Homo* that had left Africa about a million years earlier. But let's not forget about Africa itself. In Africa too, hominids were moving around and coming into contact and interbreeding. Next, Templeton discovered another migration of ancient *Homo sapiens* out of Africa at about 80,000 to 55,000 years ago. This time there was no interbreeding with the existing genus *Homo* populations that were already in Europe, Asia, and Indonesia, meaning that the ancient *Homo sapiens* then expanding its range to Europe, Asia, and Indonesia did not interbreed with *Homo neanderthalensis* or other indigenous archaic forms. This expansion is very much in line with the expansion that many lines of evidence suggest humans embarked on 60,000 years ago. One of the more important implications of Templeton's work, though, is that the recent pattern of human movement and breeding indicates a large and widespread *Homo sapiens* that has continually interbred and moved back and forth across the globe.

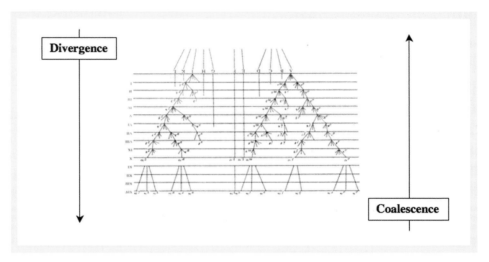

FIGURE 85. Coalescence versus divergence. The only figure in the *On The Origin of Species* is this genealogy. Darwin's description of divergence that accompanied the figure is still accurate to this day. Some population geneticists have turned the idea of divergence upside down and use the observation that in genealogies the tips of the genealogy can be traced back to a point of coalescence. The theory developed from this way of viewing divergence is called Coalescent Theory.

The 18 Eves (each letter represents an Eve A, B, C, D, E, F, G, H, I, J, K, L, M, T, U, V, W, X) and where they are found

African mtDNA haplogroups
L1   L1 : Khoi-San-speaking peoples
L2   East African : related with the Bantu expansion
L3   East African : related with the Bantu expansion
L3* Specific to Sub-Saharan Africa : present in European populations

Asian, Australian and Amerindian mtDNA haplogroups

C, D, E, G Four haplogroups represent about 77 percent of the Asian/Australian mtDNA

A, B, F Three haplogroups represent about 23 percent of the Asian/Australian mtDNA

A, B, C, D, X Amerindian mtDNA falls within these five haplogroups

M1 much greater in India than in Ethiopia.

European mtDNA haplogroups

H, I, J, K, 99 percent of all Europeans fall within these nine haplogroups

T, U, V, W and X

Around the World in 22 Million Days (Or So)

So how and when did *Homo sapiens* spread to the entire Old World and eventually to the New? Let's return to mitochondrial Eve and Y chromosomal Adam to see how male lineages and female lineages moved across Earth. We can do this at two levels. The first is at the gross level of first-range expansion into a new area. The second level is more fine-grained and can be examined to determine patterns of movement in specific regions. [Figure 86]

At the gross level, the "18 Eves" migrated in the following way: Three Eve lineages stayed in Africa. About 60,000 years ago, six Eve lineages migrated to Asia; and about 20,000 years later nine Eve lineages migrated to Europe. About 30,000 to 7,000 years ago, Eve lineages from both Asia and Europe crossed the Bering Strait and populated North and South America.

For the "Adams," the following scenario seems most accurate, based on available data: Three major Adam lineages stayed in Africa, and about 60,000 years ago seven Adam lineages migrated out of Africa and into Asia. These seven Adam lineages then migrated to Europe 40,000 years ago, and then to the Americas between 30,000 and 7,000 years ago.

From these rather simple range expansion patterns came the incredibly diverse, but extremely closely related, 18 Eve and 10 Adam modern lineages that now encompass all maternal and paternal lineages on the planet. By using more precise genetic tracers both from the nuclear genome and the mitochondrial genome, a finer-grained view of human movement can be accomplished for specific geographic regions of the planet.

Let's take a journey around Earth, starting where the first *Homo sapiens* arrived, in the Near East and Asia, about 60,000 years ago.

First Stop – Asia

Our first stop is Asia, and we will examine what the mtDNA and Y chromosomal data tell us about human movement and relatedness among all of the major Asian peoples. [Figure 87] Both mtDNA and Y chromosomal DNA data have unequivocally shown that the migrants into Asia were from Africa, and what happened after the first *Homo sapiens* entered Asia is very interesting. The genomic data indicate that the migration routes closely followed the southern parts of the Asian continent: first through Asia Minor and then into subcontinental India, followed by a migration into Southeast Asia. Once *Homo sapiens* entered Southeast Asia, there were two places to go: One major group of migrants went north into China, and another ventured into far Southeast Asia.

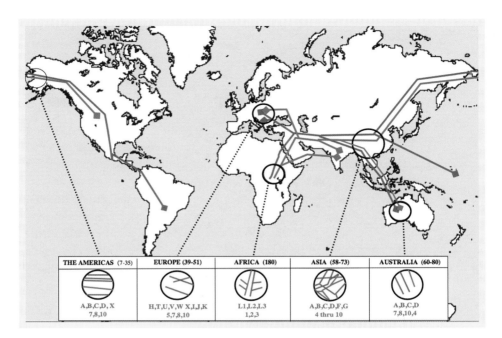

FIGURE 86. Map of the world showing the major routes of human migration as deduced from mitochondrial DNA (red) and Y chromosomal DNA (blue). The routes are much more complex than depicted in this figure. To demonstrate the potential complexity circles indicate important geographic areas where the branchings of lineages are shown magnified in the boxes below the map. The mitochondrial lineages for each geographic region are listed in red below the magnified circles. The Y chromosomal lineages for each geographic regions are listed in blue below the magnified circles. mtDNA haplotype X is most likely a European haplotype, and is also found in the Americas. The numbers in parentheses refer to possible times that the lineages entered the specified areas, in thousands of years.

Once *Homo sapiens* had established populations along this migration route, some intriguing events happened in each of the major areas of Asia. In the Middle East, an interesting story about human migration and relatedness comes from one of the most politically volatile areas on Earth: Palestine and Israel. When researchers examined the Y chromosomal DNA from many males of Palestinian and Jewish heritage in this area, they obtained an interesting result that tells a great story about the genetic closeness of humankind. [Figure 88] When they attempted to draw a branching diagram of the relationships of the various people they analyzed they obtained a web—a web that did not differentiate Palestinian Y chromosomes from Israeli Y chromosomes.

The Indian subcontinent was not just a passageway for *Homo sapiens* to the east. The routes of migration *within* this large area are also well-known from DNA sequence information. The most prominent route of migration follows the southern coast of India. The tracing of maternal lineages from people now living in India indicates strongly that three lineages both passed through and stayed in India. Using mtDNA sequences from about 800 Indian individuals and from 17 Indian tribes, researchers propose that the Indian mtDNA genes are most closely related to East Asian mtDNAs. There was also a clear disjunction between northeastern and southern Indian peoples, with the northeastern tribes

FIGURE 87. Map showing the routes of Asian migration determined from Y chromosome and mtDNA studies. Red arrows show the primary early routes. Blue lines show the secondary later routes.

being closer to east Asians than to the southern Indians. This pattern is actually congruent with linguistic information, as these latter populations also speak Tibeto-Burman languages of east Asian origin.

The Indian Y chromosome story was at one time thought to be quite boring, simply following a similar pattern of the mtDNA in its migration patterns across India. But recently geneticists discovered European or Aryan Y chromosomes in the males of Indian populations. They suggested that these Y chromosomes were introduced only about 3,000 to 4,000 years ago, probably as a result of conquest of the Indian subcontinent by European Aryan invaders. But why don't the maternal lineages give us a clue about this? Simple. If males are the ones doing the raiding and fighting, and they ventured to India in conquest without Aryan women, then there is no reason to expect that Aryan mtDNA footprints should be seen in the Indian female lineages.

Of the two lineages that split in Southeast Asia, one migrated to China, the most populous region on Earth today. Chinese scientists have taken up the task of figuring out how their ancestors migrated across eastern Asia. By analyzing 20,000 human Y chromosomes from the 58 self-described nationalities that reside in China, Chinese scientists found that the original migration of *Homo sapiens* into China occurred 30,000 years ago, and that the Yunnan and the Guangxi were the first areas to see expansion of *Homo sapiens* populations. Many thousands of years later, Chinese nationalities started to differentiate with respect to their Y chromosomes. Tibetan people are also part of this lineage. At a later period a second lineage expanded north into Central Asia. This second lineage then migrated south again to interbreed with the original lineage, whose representatives had spread northward through China. Several studies have been conducted in other parts of Asia, too.

One part of Asia we have not mentioned is Japan. Because this major landmass is an island, or a chain of them, one might think Japan would be easy to figure out with respect to the origin of its people. In an amazing show of genomics brute force (in the past it has been extremely difficult to sequence even one complete mtDNA genome), a group of Japanese geneticists sequenced almost 700 full mitochondrial genomes from people living in Japan, and added this information to an already existing 300 other Asian full mitochondrial DNA sequences. Their analysis was intended to discover the

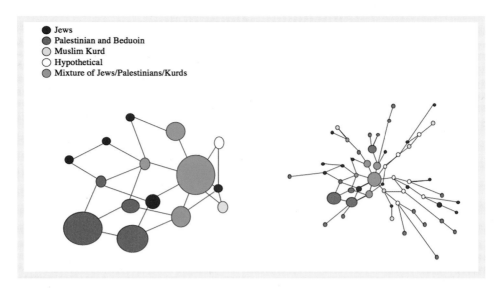

FIGURE 88. Tree of Palestinian and Israeli Y chromosomes on the right, and the enlargement on the left of the area where the various ethnic groups show networked relationships amongst Jews, Palestinians and Kurdish Muslims. Blue balls represent Y chromosomes where men self identify as Jewish, red balls represent Y chromosomes where men self identify as Palestinian, and yellow balls represent Y chromosomes where men self identify as Kurdish Muslim. The green balls represent Y chromosomes where all three ethnic groups are identified. The size of the balls represents relative frequency so that the larger the ball, the more Y chromosomes were examined. Note the interconnectedness of the various ethnic groups in the enlargement on the left.

pattern of divergence that led to the current distribution of maternal lineages in Japan. They concluded that present-day Japanese maternal lineages have arisen from other northern Asian lineages, with especially high genetic affinity to those found only in Korea. This finding is congruent with the proposed Continental connection via a land bridge to Japan after the Yayoi period (2,300 to 1,700 years ago), based on other genomic information.

This rather rapid trip through Asia has neglected one route of migration, namely the southern route *Homo sapiens* took once Southeast Asia was reached. Obviously, the next stop was Australia.

Quick Stop Australia

The island continent of Australia was colonized by the later wave of *Homo sapiens*. We know that fossil data suggest a long history of *Homo sapiens* in Australia. As we suggested in chapter 6, the Mungo 3 specimen is about 60,000 years old, indicating that *Homo sapiens* had made it to Australia at least by then, and well before this last wave of migration we are now describing (see Figure 84). But also remember that the other *Homo sapiens* for whom mtDNA analysis was possible suggest that Mungo 3 was a unique lineage that was lost or replaced by later arriving *Homo sapiens,* who got to Australia in the last wave of *Homo sapiens* migration about 30,000 years ago.

The question then becomes, where did today's Australian Aborigines come from? Neither mtDNA nor Y chromosomal DNA supports a close relatedness of Micronesian and Australian lineages, meaning that these two groups of people were founded independently of each other rather than sequentially. If Micronesian and Australian lineages were closely related, then a sequential connection of Micronesian and Australian populations could be inferred; but because there is no evidence for this notion, independent founding of these populations is a better explanation.

Specifically, the mtDNA suggests that there were several waves of colonization of Australia, but that further colonization of the continent shows very little pattern other than a north-south movement. This north-south migration must have been pretty quick, as *Homo sapiens* remains about 30,000 years old are found on Tasmania, the large island off Australia's southern coast. Y chromosomes of aboriginal Australian populations are extremely nonvariable; and, in fact, only two Y chromosomal types exist in aboriginal Australian males. This lack of diversity in male Australian aboriginal Y chromosomes indicates a recent migration of males to Australia, and a subsequent expansion of Aborigine populations. When the divergence times of these two Y chromosomes are examined, both Y chromosomes are shown to have expanded about 4,000 years ago.

If It's Tuesday, It Must Be Europe

While all of this was happening, Europe was also being explored and colonized by *Homo sapiens*. It is clear from both the DNA sequence data and the fossils that the movement of modern *Homo sapiens* into Europe had begun by about 40,000 years ago. We have already discussed the ramifications of using Neanderthal and ancient *Homo sapiens* DNA sequences to examine the multiregional versus replacement hypotheses, and found that the latter fits the facts best. But what about the specific movement of *Homo sapiens* within Europe? Because of an excellent fossil record, the basic movements of *Homo sapiens* in Europe are well known. The major factor controlling such movements during the last 30,000 to 40,000 years was probably the severe climatic fluctuations of the last Ice Age. This latest global Ice Age reached its peak 18,000 years ago, and is known for the amount of both North America and Europe that it covered with glacial material. As a consequence of the presence of ice sheets, many populations of animals, and most certainly of *Homo sapiens* as well, were broadly affected. The Mesolithic period started around 10,000 years ago, with the recession of glaciers at the end of the Ice Age. The dynamics and patterns of human genetics before, during and after the peak of the last Ice Age are critical for understanding European genome distributions.

The major question appears to be how much Paleolithic *Homo sapiens* contributed to the current European gene pool. Using mtDNA to estimate divergences, researchers conclude that there were three major phases of colonization of Europe. [Figure 89] The first occurred in the Upper Paleolithic, under 50,000 years ago. The second and third occurred at the end of the Ice Age and shortly thereafter, between 10,000 and 8,500 years ago. While these analyses detail the general timing and frequency of colonization, there are many ways *Homo sapiens* could have migrated at these three points in time. Again, mitochondrial DNA and Y chromosomes have something to say about the specifics of the colonization of Europe.

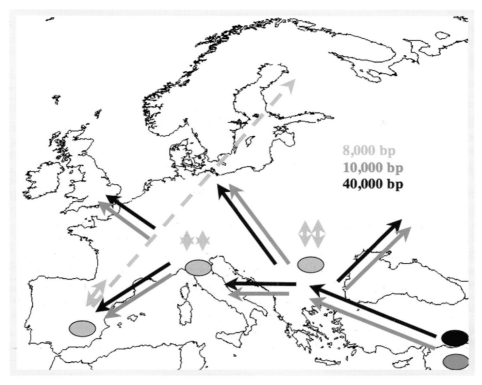

FIGURE 89. Migration routes of *Homo sapiens* in Europe in the late Paleolithic (green) and Neolithic (blue) times. Arrows represent dispersal processes inferred from the archaeological record, with the approximate date of their beginning. The yellow double arrows indicate a southward population contraction at the latest glacial maximum, followed by a northward expansion when temperatures increased; yellow ovals represent the approximate areas where these populations concentrated at the last glacial maximum (refugia). A dotted arrow represents the postglacial expansion from Iberia into northeastern Europe, proposed on genetic grounds by mtDNA information.

Two models have been proposed to explain how the European gene pool was formed. [Figure 90] The first hypothesizes that *Homo sapiens* who lived in Europe during the Ice Age expanded at the same time that post-Ice Age migrants entered Europe from the south (either from Africa or the Near East). This scenario is called the Post-Glacial Expansion. A second hypothesis suggests that the migrants who entered Europe after the Ice Age and migrated north with the recession of the glaciers made a greater genetic contribution to modern European genomes than the *Homo sapiens* already there. The picture isn't as simple as you might think though, as this second model, called the Neolithic Demic Diffusion model, suggests that as the Neolithic *Homo sapiens* migrants got farther and farther into Europe, their impact on European genomes got smaller and smaller.

One of the most interesting results of examining the genomics of European people was the discovery of a graded frequency of genes, called a cline, that changes from the southeast to the northwest of Europe. What this cline means is that European populations in the southeast have more of certain genes, and that the frequency of these genes decreases as populations are examined to the north. Most of the genes that have been

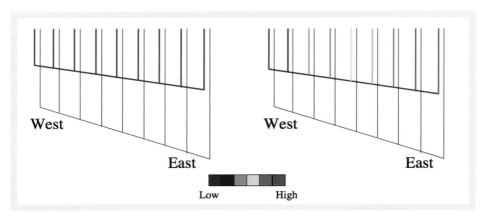

FIGURE 90. A schematic representation of the two main models of European population history: Post-Glacial Expansion (left), and Demic Diffusion (right). Past is at the bottom, present at the top. The color of the rods is proportional to the relative demographic impact of Paleolithic (black) and Neolithic (colored) immigration, under the two different models of population expansion (the NDD and PGE models discussed in the text), in several areas of Europe, along a west-east transect. Note that one model predicts a very regular consistent demographic impact, while the other model predicts an increasing demographic impact when going from west to east.

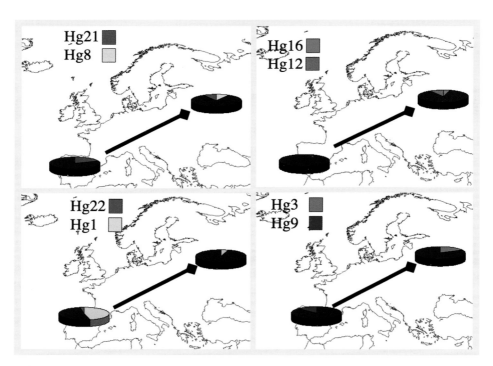

FIGURE 91. Four examples of different genes (gene designations are shown in the upper right of the figure) showing the southwest northeast changes in population frequency. The dark blue part of the pie diagram indicates the frequency of the major gene version. Note how more of the pies are filled up as one goes from south to north and from west to east. This kind of changing trend is called a cline.

examined to address this problem show this expected pattern. Figure 91 shows that Europeans in the southwest do indeed have more such genes than those in the northeast.

While it is difficult to place a figure on the contribution of the Paleolithic *Homo sapiens* vs. later immigrants, the clinal nature of most of the gene distributions examined so far does support the NDD model over the PGE model. This conclusion is reached because the NDD model suggests that there should be a higher contribution of later genomes, but that this should tail off as populations get farther and farther away from the source populations in southeast Europe.

The New World: New But Not the Newest

Many ancient *Homo sapiens* from Asia apparently discovered that the Bering Strait was an important barrier between them and a landmass of which they knew very little. The Nahuatl people of Siberia even had a word for the Western Hemisphere, "Ixachilan." There is no doubt, based on anatomical characters, that migration to Ixachilan more than likely occurred across the Bering Strait during a time when the it was either dry land or frozen over to form an ice bridge.

The New World is particularly interesting because of the large number of languages spoken by peoples in that part of the world. More than half of Earth's known dialects are from people living in North or South America, and three major kinds of languages—Amerind, Na-Dene and Eskaleut—form the base of the native languages of the Western Hemisphere. [Figure 92] The basic questions about the New World addressed by researchers using genomic information are: Where did the people who first populated the New World come from? How many times did people migrate to the New World? And what was the timing of those events?

In order to improve our understanding of events in the peopling of the New World, we first need to characterize the kinds of genomic markers that have been used. So we round up the usual suspects—the mtDNA genome and the Y chromosome—and put them to work. Five major mtDNA lineages exist in the New World. These are called A, B, C, D, and X. For the Y chromosome, researchers have used a battery of genetic markers, of which three have been most useful in addressing questions of human ancestry. These three markers are named QM3, QM242, and PM45, and they all reside on the Y chromosome.

Let's address the hypothesis that some migration occurred from the Pacific Ocean area to South America. This hypothesis was made on the basis of some linguistics and the pseudoarchaeological work of Thor Heyerdahl based on his 1947 *Kon-Tiki* raft expedition across the Pacific. Simply put, the genomic data are sparse for this scenario, some would say nonexistent. Where, then, did the migrants to the New World come from? The easy answer at the outset is Siberia. But where in Siberia? It's a really big place. Today, descendants of 31 major native ethnic groups are living in Siberia. Any one of these groups could be considered a possible candidate as the founder of the New World peoples. Among these ethnic groups, those located farthest east are the Evens, Kets, Yakuts, Yukaghirs, Buryats, Chukchi, Evenks, Altai, and, the most eastern of all, the Siberian Inuits. [Figure 93]

Researchers realized early on that just because these Siberian ethnic groups now live close to the Bering Strait doesn't mean they are the founders. Remember that there are 18 major "Eve" mitochondrial DNA lineages and 10 major "Adam" Y chromosomal DNA lineages. The assumption, using these markers on the Y chromosome and mitochondrial

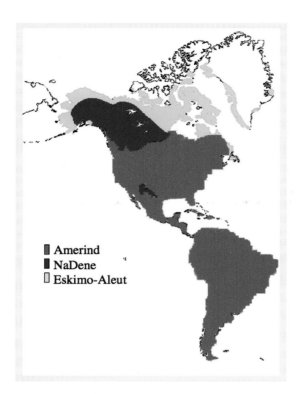

FIGURE 92. Map of the western hemisphere showing the distribution of the three major language groups in this part of the world.

DNA, is that if a Western Hemisphere group has certain markers, then by matching these markers with those present in Asian populations, a connection to the founding population can be established. The Y chromosomal data and the mtDNA data do not agree entirely, but do point to a general region of Asia that fits as a potential source for the migrants to the New World. A detailed analysis of the mtDNA data suggests that Mongolia might be the most likely suspect for the founder stock for the New World mtDNA, as all four markers—A,B,C, and D—exist in Mongolia and nowhere else in the Siberian region.

The Y-chromosomal data suggest another scenario—that two major Siberian areas contributed to present-day Native American Y chromosomes. To decipher this, scientists used the three Y chromosomal markers we discussed earlier— QM3, QM242, and PM45. It turns out that the majority of the Y markers in Siberia are PM45 and QM3, both of which can be traced back to southern-middle Siberia. More specifically, these two markers suggest that the Kets and the people from Altai are the groups most likely to have given rise to the Native American Y chromosomes – but to only two of the existing major groups, the Amerinds and the Na-Dene. The best we can say in light of these somewhat conflicting data sets is that the general area of southern-middle Siberia are the source for the majority of New World natives Y chromosomes and mtDNA.

The first thing that comes to mind when we think about the crossing of the Bering Strait by founders of the people of the New World, in particular the Na-Dene and the Amerinds, is, how many people crossed at initially? Scientists have used the coalescent approaches described earlier to estimate how many individuals were in this founding wave that gave rise to the Amerinds. This is because Coalescent Theory takes into consideration all those demographic factors we described, such as migration rate, mutation rate, and, most importantly for our question of how many people crossed the Bering Strait, population size. This research arrived at the stunning estimate of 80 people who "walked" across the Bering Strait. While this number might seem small to some readers, one thing for certain is the size of the founding population was less than a few hundred.

But wait! How about the third major language group, the Eskaleut? One way to approach their origins is to ask how many waves of migration there were. The Amerind/Na-Dene wave might have occurred before or after the Eskaleut wave. Again, mtDNA and Y chromosomes have a tale to tell. Some scholars suggest that there were three major waves of migration to the New World that correspond nicely with the three major Native

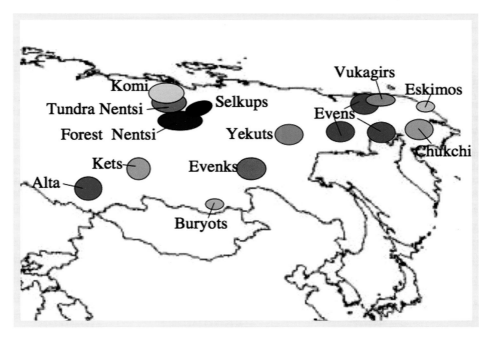

FIGURE 93. Map of Siberia showing the locations of the major tribal groups that reside in the area.

American language groups—Amerind, Na-Dene, and Eskaleut. According to this theory, a very small number of Amerind ancestors, estimated at fewer than a couple hundred, arrived about 20,000 to 15,000 years ago, followed by the Na-Dene ancestors at about 10,000 years ago, and then by the Eskaleuts ancestors, who arrived in Alaska and northern Canada within the last 7,000 years. Some earlier mitochondrial DNA research supports this theory. This work found that all Native American's mtDNA comes from four mtDNA lineages (A through D) and that, significantly, Amerinds have all four lineages, Na-Dene only A, and Eskaleuts A and D—suggesting different migrations at different times.

Other scholars favor a two-wave scenario. Here, the initial entry date of 15,000 years or so coincides with a lot of archaeological data, and supports a pre-Clovis entry into the New World. In addition, the time of the first entry into the New World lies smack in the middle of the time the latest glaciers were at their maximum. Because it is highly unlikely the migrants walked across the glaciers, they almost certainly followed a coastal route, where the glaciers would not have impeded their progress.

The remainder of the New World migrants came in a second wave. This migration can be explained by going back and remembering that five mtDNA haplotypes are found in Native Americans – A, B, C, D, and X. The X mtDNA haplotype most likely entered the New World long after the A through D markers initially did. The estimated date for the entry of the X marker is about 12,500 years ago, and coincides with the recession of the glaciers in North America. Because the X mtDNA marker was around at a time that favored its spread, it has been incorporated into many Native American populations in the Northern Hemisphere, showing that sometimes coming late to the party is a good thing to do.

By examining the presence or absence of the major mtDNA markers that exist within the three major language groups in Native American populations, we can also get some

idea as to how these groups spread to populate the entire Western Hemisphere. We can look closely at the situation in Amerind-speaking people as an example. Scholars have discovered that Amerind-speaking people have different frequencies of these markers in different parts of the Western Hemisphere. Amerind speakers have the four markers—A, B, C, and D—in most populations, but A decreases from north to south. Furthermore, markers C and D increase from north to south, while B is mostly constant but is completely absent in the far North. The mtDNA marker called X is found exclusively in North America. Because there is a concentration of these markers in North America, and clines exist from north to south, we can conclude that North America was the first major stopping-off point for all three groups of Native Americans, and that there was a gradual and directed migration of people from north to south. This makes sense, but it is nice to see the majority of the

FIGURE 94. A well-studied culture named after the Clovis archaeological site in New Mexico, marks the turning point in the history of Western Hemisphere colonization. Clovis artifacts are known from several sites dating between 11,500 and 11,000 years ago.

archaeological data and genomic data agree with this intuitive pattern of movement of the people who first came to the Western Hemisphere. [Figure 94]

The Pacific Ocean: The Final (Earthly) Frontier

Our next stop is the Pacific Ocean region and the islands of Remote and Near Oceania in the large Oceanic areas of Polynesia and Micronesia. These migrations of *Homo sapiens* to the islands in the Pacific Ocean happened within the last 10,000 years and represent the last major prehistoric human dispersal event before the very recent occupations of New Zealand and Madagascar. The major question in the peopling of Oceania is, of course, "who did it?" Linguistic analysis suggests that the source population might be from Taiwan. Analysis of mtDNA from Oceanic peoples is in line with this suggestion, but also cannot rule out origins in Island Southeast Asia (a triangle formed by Taiwan, Sumatra and Timor). Y chromosomes tell a very different story though, suggesting that the Taiwan colonization was completely independent from the colonization of Oceania and that there is no substantial contribution of Melanesian Y haplotypes to Polynesian populations. An interesting analysis of mitochondrial DNA indicates that the island of Ponope (in the Micronesian Caroline Island chain) might have been a hub of activity for interbreeding and migration of people from Remote and Near Oceania.

Africa: Meanwhile, Back at the Ranch

Back in Africa, of course, lineages were also migrating. Because this is where it all started, the Eve lineage in Africa is much deeper than in other places, reaching back to ages of 100,000 years and more. This Eve lineage is designated "L," and in Africa there are now four major subdivisions of this L lineage. The major trick with these lineages has been to match them with the major language groups in Africa. One of the lineages correlates with the Khoi-San language groups and two others with the Bantu-speaking peoples. A fourth lineage seems to parallel the movement of sub-Saharan African people and is also found in almost all European populations, most prominently in Finns. In fact, one of the two Bantu mtDNA haplotypes is also found in low frequencies in Nordic peoples.

Timing of the movement of these lineages has also been possible. The first lineage that expanded was the Khoi-San, probably starting 150,000 to 100,000 years ago. This lineage did not leave Africa. About 60,000 years ago the two Bantu lineages started to expand from east Africa to the rest of the continent.

Specific attempts to correlate genome markers with other indicators of the movements of people have resulted in some interesting inferences. One, in particular, involves the establishment and movement of people using "clicking" consonants in their speech. About 30 languages in southern Africa use click consonants. One is spoken by the San. The Hadzabe people of eastern Africa also use click consonants, but do not appear to be at all closely related to the San. Genetic analysis suggests that these two groups are more distinct from each other than any other pair of African populations, indicating an extremely deep branching between them. The implication is that the click consonants have existed in African languages for tens of thousands of years, and are not simply recent acquisitions into these languages. One of the more imaginative inferences from these results is that clicks may have been retained in these languages because they confer communication advantages during hunting.

Hitching a Ride

While we have emphasized the use of human mtDNA and Y chromosomal markers as being useful in understanding the movement of people across Earth, we can also use other organisms that hitch rides either on or in our bodies or that travel with us as markers of our movement. What kinds of organisms are these? Well, think about the kinds of things that get on or in us. Fleas and viruses immediately come to mind as pests that have hitched rides on humans.

Several viral species have also been used to examine the movements of people. One example is the hepatitis G virus, an RNA virus identified only in the last decade.

Because it does not cause liver damage like its relatives hepatitis C and B, it appears to happily coexist with humans and is not eliminated from human populations. It has therefore been suggested as a good marker to follow human population movement, much as human Y and mtDNA are used. One recent study examined nucleotide sequences of this virus in several hundred humans of diverse geographic origin, and showed that the virus probably originated in African people and then was transmitted via migrating people to Southeast Asia, with viruses from European people less closely related. These results are in line with the mtDNA and Y chromosome stories discussed previously. Other viruses where this approach has been useful are the human polyomaviruses, JC viruses, and the T-cell lymphotropic viruses.

In addition, the bacterium that is responsible for some stomach ulcers, *Helicobacter pylori,* has also been used to track human migrations. These pathogens tell us stories very similar to the mitochondrial DNA and Y chromosomal DNA markers. For instance, the ulcer-causing bacteria in South American natives can be traced to the same strains in eastern Asian populations and not in Europeans, corroborating the close affiliation of Amerindian people with east Asians that was demonstrated with mitochondrial DNA and Y chromosomal DNA.

Some organisms, such as rodents, hitchhike in the vessels or the supplies that people take when migrating. In particular, rats and mice are good sources of information for migratory patterns of humans. In one study researchers examined the genetic relatedness of rats to examine the colonization of Oceania by humans. We are all probably familiar with the rats common in Europe called *Rattus norvegicus*. These rats hitched rides to the New World during the European exploration phase of this hemisphere. The rat used in the Oceania studies is a cousin of *norvegicus* that looks a lot like this European rat and is called *Rattus exulans*. [Figure 95] Because this rat is genetically distinct from the European or Norway rat, the more recent movement of Europeans can be distinguished from the ancient movement of the first settlers of Oceania. Most importantly, this rat cannot swim, so its movement has to be facilitated

FIGURE 95. Carving of rats (upper right) in a Polynesian figure from an ocean-going vessel. Copyright Tim Mackrell/PNAS.

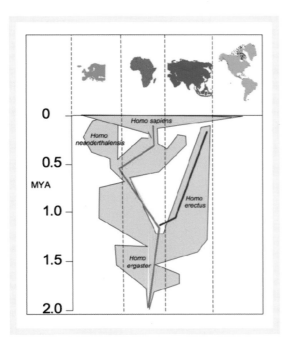

FIGURE 96. Diagram showing the divergence of genus *Homo* species from a *Homo ergaster*-like ancestor. The split between *Homo sapiens* and *Homo erectus* occurs about at the same time the two *Pediculus humanus* lineages split. The red line indicates the trajectory of the *Homo sapiens*-specific lice lineage and the blue line the other potential *Homo* lineage. The point where the blue lineage jumps into the *sapiens* lineage is the proposed point of physical contact for the transmission of the other lice lineage to *sapiens*.

by humans. On many Pacific Islands, these rats are found both alive and as fossils. DNA can be isolated from both the living and fossil rats from a wide variety of islands, and used to look at how people might have moved around in the Pacific. The bottom line is that while the general patterns of movement we have gleaned from human mtDNA and Y chromosomal DNA are accurate, the nitty-gritty details of how movement occurred among islands is very complex and involves much more cultural interaction of people in this area than was at first thought.

Lice provide our last story of human movement. Because lice are found on other primates, their distribution on different organisms can tell us something about even older human movements than viruses and rodents can. Lice of two lineages live on genus *Homo*. One of them is the species *Pediculus humanus*, the human head louse. Analysis of these lice indicates that this lineage split into two major lineages about 1.2 million years ago. This information means that, when this louse lineage split happened, one lineage stayed with the lineage leading to *Homo sapiens*, and one went with another lineage in the genus *Homo*. [Figure 96] The kicker is that both of these lice lineages now live on us.

One underwent a bottleneck event about 100,000 years ago and is thought to be the lice that have lived on *Homo sapiens* and most of the ancestors of our human lineage all the way back to 1.2 million years ago. The other lice lineage found on humans is thought to have been acquired subsequently. The study suggests an interesting series of events in the evolution of the genus *Homo*. The most likely way this second lice lineage could have been able to infest our ancestors is through direct contact, suggesting that at some time our *Homo sapiens* ancestors were in direct contact with another species of *Homo*. [Figure 106]

X Marks the Spot

We have now completed our maternal and paternal journeys around the world. The two tools we have used over and over again tell us specifically about maternal and paternal patterns of movement and evolution. But what about how male and female lineages have mixed over the past 60,000 years? Our best bet for understanding this mixing

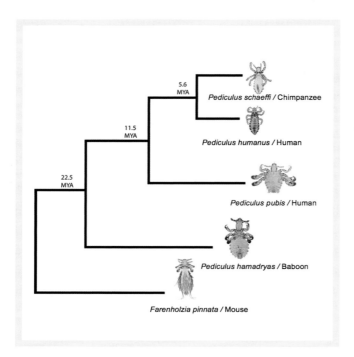

5.6
MYA

Pediculus schaeffi / Chimpanzee

11.5
MYA

Pediculus humanus / Human

22.5
MYA

Pediculus pubis / Human

Pediculus hamadryas / Baboon

Farenholzia pinnata / Mouse

FIGURE 97. Lice phylogeny. Numbers above the branches indicate divergence time from present. Numbers below the lines are measures of confidence in the branching order (numbers closer to 100 are better). Species names are italicized and their primate hosts follow.

is to look at our third genetic tool—the X chromosome. The X chromosome tells us something really different about how humans have evolved during the last 60,000 years.

Using a large stretch of the X chromosome called Xq13.3, from several human genomes, scientists discovered the following pattern of relationships among them. [Figure 97]

There are two really important results that arise from this X chromosomal tree. The first is that the tree represents a potential African origin for this genomic region, as we saw for mitochondrial and Y chromosomal DNA. The arrow in the tree represents where the tree would be connected to the common ancestor of all X chromosomes. Note that this arrow is pointing to an African X chromosomal type. In fact, of the 10 genomic regions with sufficient numbers of humans assayed for the genomic region (ß-globin, dys44, Gk, MC1R, mtDNA, PDHA1, PLP, Xq13.3, Y chromosome, ZFX), nine of them are consistent with an African origin. This result is nice corroborating evidence that all human X chromosomes come from Africa, too. The second result is that the X chromosomal DNA tree indicates that, as we begin to look at non-maternal and non-paternal lineages, the hierarchical relationships among individuals start to blur. In other words, the typical neat branching order of mitochondrial and Y chromosomal DNA breaks down, and the relationships of humans from different geographical areas begin to look more and more like a web.

Why is this, and what does it mean? First the why. As we pointed out earlier, unlike Y chromosomes and mitochondrial DNA, an X chromosome can recombine with other X's in female reproductive cells. Recombination of the X chromosomes mixes up all of the information on the X and starts to mix up the genealogical signal, as people interbreed more and more with each other.

Now for the what. The webby appearance of the X chromosome tree is the direct consequence of the ability of people of different continental origins to interbreed with each other. It is a story that our autosomes will tell us too, because, just like X chromosomes, the autosomes are prone to recombination effects. This view of human relatedness also means that, as we look at more and more of the genomes of different people and attempt to use our whole genomes as markers for our ancestry, we will get the answer that all

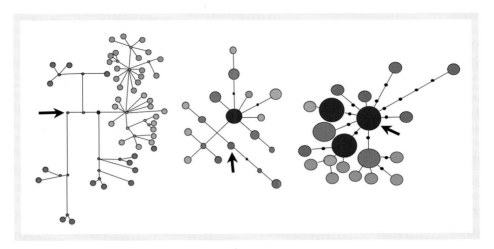

FIGURE 98. Phylogenetic trees showing the relationships of various human genes for mtDNA (left), a Y chromosomal gene (middle) and a large stretch of DNA from the X chromosome. Red balls are genes from people from the continent of Africa. Green balls are genes from people Europe and orange balls are genes from people the continent of Asia. Purple balls indicate genes that are from people from two continents and blue balls represent genes from people from three continents. Note that people from the North American continent are not in this figure.

humans are a part of a large web-like structure united by our common humanity. This view nicely explains why there is more variation within human geographical groups than between them, and why any two randomly chosen individuals will differ at only 0.1 percent of their DNA sequence positions.

CHAPTER EIGHT

THE BRAIN–THE KEY

All Roads Lead to ...

Throughout this book, we have emphasized our links with our planet and with the rest of the organisms on Earth. We have highlighted the connections humans have with our close primate relatives, and our even closer relationships to the other members of Hominidae and the genus *Homo* that have lived on this planet. It is now time to look at where we differ from other organisms.

How does the fossil record help us understand our differences from other organisms? What do our genomes tell us about our uniqueness? What is it that makes us specifically human, and in the process makes us different from all other organisms on the planet? It is a pretty good bet that we are the only species on this planet that cares what the answers to such questions are—or can even ask them. And this, of course, is part of what makes us unique. But why?

Our genomes have the same kinds of genes that other organisms on this planet have. Our genomes are organized just like many other organisms' genomes. Our cells work in exactly the same way as those of all other organisms and follow the basic rules of molecular biology. Our tissues look very similar to other species' tissues, even to the point where we can accept transplants from other species. Our bones are not spectacularly different from other organisms in any way.

Yet the fossil record tells us that something very important happened in the events leading up to *Homo sapiens*. The fossil and archaeological evidence gives us comparative information on how our predecessors' anatomies, behaviors and cultures changed over time. Where did these changes happen and how? When we ask this question, all roads lead ultimately to the human brain and to the many new behaviors opened up by its peculiar evolutionary history.

But what is it about our brains that makes us unique? One way of looking at this question is to start not with the brain itself but its products: our behaviors. The author Don Brown once listed the things that make us unique as humans—the so-called "List of Human Universals." Here it is, quite a long laundry list of our evolutionary legacy:

> Abstraction in speech and thought, actions under self-control distinguished from those not under control, aesthetics, affection expressed and felt, ambivalence, anthropo-morphization, ART, baby talk, belief in the supernatural/religion, beliefs about death, binary cognitive decisions, body adornment, childbirth customs, childhood fear of

strangers, choice making (choosing alternatives), classification (of behaviors, inner states, weather, conditions, TOOLS), collective identities, conflict, conjectural reasoning, containers, cooking, cooperation, copulation normally conducted in privacy, coyness display, crying, culture, customary greetings, DANCE, daily routines, death rituals, decision-making, distinguishing right and wrong, division of labor by sex, dreams, dream interpretation, emotions, empathy, envy, etiquette, explanation, facial expressions, family, fears, figurative SPEECH, fire, folklore, food-sharing, attempts to predict the future, generosity admired, gift-giving, distinguishing between good and bad, gossip, government, GRAMMAR, hairstyles, healing, hospitality, hygiene, in group and out group, insulting, interest in living things that resemble us, jokes, kinship statuses, LANGUAGE, law, leaders, logic, lying, magic, marriage, materialism, meal times, meaning, measuring, medicine, memory, mood- or consciousness-altering techniques, mourning, murder proscribed, MUSIC, myths, NARRATIVE, overestimating objectivity of thought, pain, past/present/future, person (concept of), personal names, planning, play, POETRY, possessiveness, practice to improve skills, private inner life, psychological defense mechanisms, rape, rape proscribed, reciprocal exchanges (of labor, goods or services), RHYTHM, right-handedness (mainly), rites of passage, rituals, sanctions for crimes, sense of self (distinguished from others and responsible), sex (attraction, jealousy, modesty, regulation), shelter, social structure, status and roles, sweets preferred, SYMBOLISM, taboos, time, TOOLS, triangular awareness, true and false distinguished, turn-trying, tying material (string, etc.), violence, visiting, weapons, world view.

Although Brown seems to have been trying a bit too hard and to have included in his list some activities that are specifically human only arguably at best, we are sure that every reader of this book will be able to make several additions to this list of things that make us human. And indeed, any list of things that only humans do could be prolonged almost indefinitely (Brown says nothing about doing crossword puzzles, for instance, or hang-gliding).

In the end, simply adding to the list gets us nowhere, because all of the items on it can be traced back to a more generalized capacity that makes all of them possible. It is this capacity that we need to grasp. Still, if we are to understand exactly how we differ from our closest relatives, we do need to have some characteristic behaviors to focus on. For notable as our anatomical differences are from bonobos and chimpanzees, it is our behaviors that present us with an apparently unbridgeable gap between us and them. We have capitalized a few traits in Brown's list that we think are particularly worthy of attention, and we'll look more closely at them later on, but this chapter is less about behavior than about what makes behavior possible. It's no secret that the enabler of all these things is the brain. That means, of course, that in order to understand what makes humans truly unique, we need to know something about that mysterious organ that resides within our unusually globular heads.

What Is a Brain, Anyway?

What is a brain and what is it made of? While what the brain is might be intuitively obvious to many readers, some important issues need to be addressed to clarify this question. To achieve this clarification, we need to understand both what the brain does and how it is constructed. Exactly what goes on inside the brain to produce what we experience as consciousness remains mysterious after well over a century of research, but we can look at how our brains are built with its extraordinary products in mind.

From the strictly physical aspect, brains are simply tissues – extremely complex tissues, but tissues nevertheless, made of cells containing genes and working with the same general kinds of cellular processes seen in other tissues. What makes the brain and the nervous system (which we will consider as simply an extension of the brain) different from the rest of the tissues in an organism is the unique ways in which brain and nervous tissues go about their various businesses despite sharing the same basic molecular processes with other tissues. For instance, genes are basically expressed in the same way in brain as in other tissues, just as proteins are processed and utilized similarly. The difference lies in the kinds of genes that are expressed, and in the functions of the proteins in the brain. Fortunately, the basic similarity in the way all genes and proteins work makes it feasible to study the function and makeup of the brain.

While creatures other than animals obviously do not have brains as we understand them, they can sense their surroundings. So do they have brains in some sense? Even if they don't, they are still able to help us define what a brain is. For while bacteria and other single-celled organisms are just that, single-celled, and thus can't have brains or even nerve cells, they do have unique biochemical systems that can sense their surroundings and direct their genomes to respond to the environment. For instance, in bacteria there is a process called "quorum sensing" that uses specific kinds of microbe-specific molecular processes to sense the presence of chemicals in the environment (hence, the presence of other microbes), and directs the bacteria to respond by expressing a protein or a batch of proteins. This quorum-sensing mechanism has been essential for the survival of some bacteria, as it allows them to sense their surroundings and rapidly respond to challenges from the environment.

Plants form another category of organisms that do not have nervous tissue as classically defined, yet can sense their surroundings and respond. Surprisingly, plants actually produce some of the same proteins that organisms with brains do. In particular, proteins known as "glutamate receptors" behave in plants in the very same way that they behave in animal brains. In animals, glutamate and other chemicals are secreted from nerve cells as a means of communication among nervous tissues. Chemicals like these are called "neurotransmitters," because they help transmit information among cells in nervous tissues like the brain. Embedded in the outer layers of nerve cells are proteins that receive neurotransmitters such as glutamate.

When the glutamate receptors encounter glutamates in an animal they regulate the intake of the chemical into the cell, and the buildup of the glutamate in the cell then trigger another response that is carried along to the brain. In animals, faulty glutamate reception and transmission result in a lack of neural response among brain and nerve cells. In fact, glutamate overload has been linked (post mortem) to schizophrenia, and inefficient signaling via glutamate may also be involved in the manifestation of Alzheimer's disease. So glutamate is an incredibly important molecule in the animal brain. But are there plants with Alzheimer's disease? Could there be schizophrenic plants? Probably not, but plants with glutamate receptor defects have an altered ability to do certain things correctly, just as faulty glutamate reception causes neurological defects in Alzheimer's disease and schizophrenia. When compounds that block glutamate reception are fed to experimental plants called *Arabidopsis thaliana* (a small flowering plant also called thale cress), their stems grow really long and they begin to look pale compared with other plants because the production of chlorophyll, the protein pigment that makes plants green, is reduced.

Why? Eliminating glutamate reception blocks the ability of the plant to respond to light, which in turn reduces chlorophyll production. While it would be difficult to demonstrate, but intriguing nevertheless, that this response to blocked glutamate receptors in plants is the same as what we see in Alzheimer's or schizophrenia, it *is* reasonable to conclude that plants and animals have the same ancestral system for communicating among cells. In plants the system evidently stayed quite primitive, but in animals the system evolved to be an important component in how nerve cells communicate one with another. The story doesn't stop there, though. In the primitive plants known as cycads a glutamate receptor blocker is produced that wards off insects and other animals. The cycads are able to discourage predation in this way because ingestion of cycad tissue causes animals to lose memory and act demented. This goes for humans too: In the Solomon Islands, natives who eat cycad parts have often been stricken with extreme dementia. This connection of plant glutamate reception with animal glutamate receptors is yet another interesting link in the tree of life, and a great example of how two very different lineages of organisms—plants and animals—have retained the ancestral function of the same protein, even as in one lineage the protein has been pirated to participate in an even more complex role in the brain.

Moving on to animal brains, we can ask whether the so-called "lower animals" are brainless. It turns out that sponges and Placozoa (remember the small three-cell layered organisms that are on the lineage connected to the most ancestral animal) are the only lower animals that do not have brains. Yet sponges probably have glutamate receptors and other cell-to-cell communication systems that serve to communicate information between cells. In evidence of this, if you touch the outside of a sponge its body will contract, indicating that some form of communication among cells is at work. The next-lowest group of animals, called cnidarians (medusa, jellyfish and other squishy things), has what is called a primitive nerve net. An animal of this kind, known as a hydroid, has several cells that form a cell network, allowing communication along the entire length of the body.

Animals used for experiments, such as worms and flies, have interesting histories of study of their nervous systems. The group of worms known as nematodes was specifically chosen as a model organism because of how easy it is to do genetics and developmental biology with it. These characteristics make it a great animal for coming to an understanding of its nervous system. The species *Caenorhabditis elegans* is a preferred experimental subject because the number of cells in its body is small (959), and because biologists understand where every cell comes from and goes to during the development of the worm. While most of the 959 cells are dedicated to reproduction, nearly one-third of the total cells of the body of *Caenorhabditis* are neural cells—so while there is no brain to speak of, there is lots of nervous tissue. And the nerve tissues are fairly well-organized and not simply net-like as in the Cnidaria. The nematode head, for instance, is packed with sensory organs or nerve cell clusters that respond to light (though it has no eyes), smell, taste, touch and temperature. Nematodes can even remember things.

As we will see when we look at other model organisms later in this chapter, memory in nematodes is accomplished much in the same way as it is in most organisms that have memory.

Fruit flies have an even more complex nervous system, and they also have brains. The fruit fly brain is complex and services the fly through an even more elaborate nervous system that spreads across its entire body. Developing flies start with about 85 precursor or stem cells for their brains, and end up with over 10,000 cells that carry out nervous system functions. These cells belonging to the nervous system are called neurons. Memory has also been a big-ticket item in the study of fly nervous systems. Fly biologists have done large screens for flies that have bad memories. In essence, they look for really stupid flies. Fly biologists often fancy themselves as comedians, manifested in the way they name things. The flies they have found with memory problems possess one of the more imaginative sets of names in biology. For instance, two mutants that are relatively —actually really, really—stupid were named for vegetables: "rutabaga" and "turnip" (a third bad memory gene mutant fly just as stupid was just called "dunce" when its namers ran out of vegetables for names). All of these mutants were produced in the lab and have been used in studies aimed at characterizing the role of genes in memory.

As one starts to examine vertebrate brains, one sees that more and more cells are dedicated to neural tissue as brains get bigger. One vertebrate in particular, the mouse, is easy to breed and to manipulate experimentally, so it has been a major subject of brain researchers. The reason this approach works is that the mouse brain, along with all other mammalian brains, has the same general structure as a human brain. Mammalian brains are, in general, scaled to the size of the animal. [Figure 99]

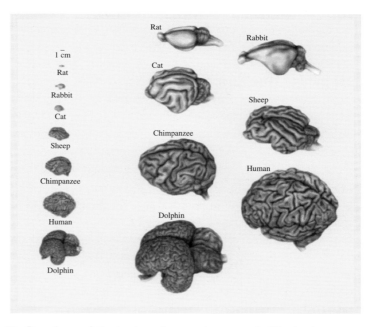

FIGURE 99. Drawings of the brains of several mammals. The brains are arranged by size on the left, and enlarged drawings are shown labeled on the right.

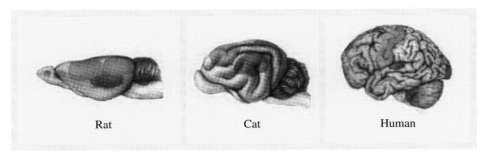

| Rat | Cat | Human |

FIGURE 100. Drawings of rat, cat and human brains (left to right). Color shaded regions refer to corresponding regions in the three different animals. This figure shows the growth in relative size and importance of the associative areas of the brain, from rats to cats to humans. Green = sensorimotor center, red = visual center, blue = auditory center.

Structures in the brain are also fairly well-conserved. What this conservation means is that certain functions of the brain are always found in the same place in the brains of different animals. [Figure 100] More importantly for our understanding of nervous system function, studies of mammalian brains have revealed that the various regions of the brain that accomplish certain jobs like olfaction, sight and other neural functions are conserved in the same general regions of the brains of other mammals. It is one thing to have a similarly shaped and functioning brain to humans, as most mammals do, but quite another to unlock the secrets of these brains. Similar studies on nonhuman animals have determined regions of the brain that serve other functions, and in mammals it is evident that the location of certain functions in particular regions of the brain is well-conserved. One interesting observation is that while mammals tend to conserve the general location of centers for certain functions (e.g., sensorimotor, auditory and visual centers), as organisms become generally more "intelligent" we find more neural real estate, presumably opening up new areas of brains for novel or expanded functions.

Frankenstein vs. the Sea Slug

Brain structure in and of itself tells us little about function, and function is obviously what we are after when we look at the brain as a key determinant of what makes us human. One approach that has aided scientists in understanding brain function is the judicious choice of model organisms, as we have discussed above. We have mentioned several choices made by scientists so far such as the worm, because it has a simple nervous system structure, and the fruit fly, because it can so easily be genetically manipulated. One model organism we haven't mentioned but that adds greatly to our understanding of how the brain and nerves work is the lowly sea slug (*Aplysia*), made famous by Eric Kandel, one of the 2000 Nobel Prize winners and University Professor at Columbia University. His work took advantage of the unique nervous system of the sea slug to determine the molecular and physiological basis of memory. The sea slug used by Kandel has only about 10,000 neuronal cells in its body. It also has a very simple reflex action, called gill retraction, that can be used to study memory. When a part of the sea slug's body, such as its mantle shelf or siphon, is touched, the sea slug will withdraw its gills. By studying this reflex action, Kandel and his colleagues were able to pin down the pathways of response to external stimulation in the nervous system of the sea slug. More importantly, they were able to deduce the biochemical pathways that were involved in the response to external stimulation and in the creation of memory in the sea slug. One of the most interesting aspects of Kandel's work is the discovery that neural connections, that is, how nerve cells are connected to each other in the brain, are an important aspect of how the brain works. Kandel has further shown that cellular and molecular fine-tuning of these neural connections is an important part of memory and presumably of other kinds of brain function.

With *Aplysia*, *Drosophila* and *C. elegans*, the nervous system and certain individual functions like learning can be analyzed in detail. These and other model systems have given us an unprecedented view of neuronal activity that appears to be well-conserved across a wide range of life on this planet. Because the fine-tuning of connections is a molecular, genetic and developmental process, Kandel has suggested that there is a strong link between cognitive psychology and molecular genetics. In other words cognition can, in many ways, be reduced to molecular processes. Kandel believes that this connection facilitates a new science of mind and may demystify many of the strange things our human brains do.

The study of human brain function is one of the most interesting yet difficult things to accomplish in science. There are more than 10 billion cells in the human brain, each and every one making connections with others. The complexity of the structure of this mass of tissue in human beings seems overwhelming at times. But over the past century or so, scientists have found ways to tease apart the way the human brain works. They had to be very careful, though, because of social taboos on experimenting with the human body. While early Renaissance anatomists such as Andreas Vesalius and William Harvey were able to make important advances, much research on the human body was not readily accepted. We all remember the villagers chasing Frankenstein along cobblestone streets—a metaphor for society's early aversion to probing the human organism too deeply. Indeed, it wasn't until the latter half of the 19th century that the scientific value of anatomical work on the human body finally became fully accepted. Many anatomists in England and France during the 18th and early 19th centuries were reduced to dealing with grave robbers and murderers to obtain material to advance their knowledge of human anatomy.

Experimentation on, and visualization of, human brains could not be accomplished readily. Later, even when genetics had become an important part of modern biology, studying the human brain from a genetic perspective remained tricky and difficult because scientists just couldn't do breeding experiments on humans the way they did, for example, with plants. To compensate for this, scientists developed methods for observing the structure of the brain at the cellular and tissue levels, without violating the taboos of scientific research on humans. One approach took advantage of the more relaxed attitudes of society in the latter part of the 19th century that allowed scientists to examine the remains of humans post mortem. Scientists could observe neurological or behavioral disorders of people during the lifetimes of patients, and then were able to autopsy the individuals after their death. In this way the brain of a person with a memory or speech disorder could be examined after the person had died to determine if and where a lesion had occurred that might have been responsible for the behavioral or neurological defect. Let's look at three major regions of the brain as examples of human-specific brain function characterized in this way – Broca's region and Wernicke's region for language, and the region of the brain responsible for memory.

The 19th-century neurologists Paul Broca and Carl Wernicke observed people with language disorders. Broca studied people with disorders in the ability to speak or comprehend language due to brain injury called aphasias. In particular, the people he studied were unable to sustain articulate speech for more than a few words. They could answer "yes" and no, but were completely incapable of speaking in complex sentences. Wernicke, on the other hand, studied people who were incapable of discerning what was spoken *to* them. Another kind of aphasia was specific to *hearing* spoken language, for the patients with this aphasia were able to read and comprehend the written word. Upon their deaths, the brains of both sets of patients were examined for any abnormalities in structure, and both researchers noticed areas of the brains of their subjects that consistently showed lesions.

In Broca's case, the lesions were in a part of the brain called the frontal lobe of the neocortex; in Wernicke's case, they were in the temporal lobe. In recent years, researchers have become less confident of the localization of the deficits to the exact regions identified by Broca and Wernicke, and the precise boundaries of the brain regions named after these two researchers are still debated. Despite these problems, the general method of localizing brain functions established by Broca and Wernicke is still used by neuroscientists today. The main difference is that modern methods use brain imaging on live subjects, which means that researchers don't need to wait for their subjects to die.

Memories

More difficult to pin down than those regions associated with speech were the regions of the brain responsible for memory. While treating epileptics, a Canadian neurosurgeon named Wilder G. Penfield in the 1920s and 1930s used electrical stimulation to map several kinds of functions in the brain. Whenever he had a patient who needed surgery, he also obtained permission to do some electrical probing of the surface of their brains. Surprisingly enough, even though our brains process all the information from our bodies about outside stimuli including pain, they have no pain receptors themselves. What this means is that a patient undergoing brain surgery can experience the probing of his or her own brain without having to be under general anesthesia. Hence, the surface of a patient's brain could be probed during an operation; the patient could tell the surgeon his or her response to the electrical stimulation, and no pain would be experienced. In this way, during his treatment of epileptics Penfield performed more than 1,000 experiments in which he probed most of the surface of the brain by electrical shock. As these experiments proceeded, Penfield noticed that he could produce a memory flashback when he stimulated certain parts of the brain. The parts of the brain most likely to produce this response were the temporal lobes.

Later, also at McGill University in Montreal, Canada, another researcher named Brenda Milner combined the neurosurgical approach with the classical approach of Broca and Wernicke. A patient called HM with normal capacity for memory (patients are almost always given pseudonyms or identified by their initials), but who was suffering from severe epileptic seizures, underwent surgery to correct the seizures. The surgery involved removal of part of the temporal lobes. While HM's epileptic seizures decreased in number and intensity after surgery, HM developed severe problems in the memory department. HM developed an early version of the "40 first dates" syndrome that Drew

Barrymore made famous in her movie about a young woman who was in a car crash that damaged her temporal lobes. In the movie, each day Barrymore's character wakes up, she remembers only what happened to her before the accident. Her brain is incapable of storing memories from the day before. The 40 first dates of the movie's title refer to the efforts of Adam Sandler, who has each day to court her anew because her brain cannot store the short-term memory of the day before and convert it into long-term memory.

After the temporal lobe surgery to correct his epileptic seizures, HM had the same problems of conversion of short-term memory to long-term memory. But in both cases, motor skills were not affected. In other words, things like sensitization, reflex response and conditioning (responding to a stimulus in a certain way because memory is involved) *could* be stored in the long term. And while we don't know how HM fared for the rest of his life, there was a fairy tale ending for Barrymore and Sandler after their 40 first dates that didn't involve a cure for her inability to store short-term memory and convert it to long term memory.

Picturing Brains

The gross structure of the brain is well-known, and a few major functions of the brain are easily mapped to specific regions of that mass that sits in our skulls. But the precise localization of certain functions of the brain has been more difficult. The idea behind getting an image for a function of the brain is to provide an outside stimulus to an individual and to look at what part of the brain responds. One of the early techniques for viewing the structure and function of human brains was called pneumoencephalography. This method of viewing brains was even more brutal than its name and entailed the approach of replacing the skull's cerebrospinal fluid with air, followed by X-raying the poor subject's head. This approach, used in the early 20th century, made the brain images clearer on the X-ray. But the obvious danger of brain damage by being made into an airhead led to its replacement by angiography, involving the injection of dyes into the bloodstream that could then be followed using radiography.

These early methods gave a very static view of the brain, with more detail of structure than of function, but in the 1970s computerized axial tomography (or CAT) scanning advanced imaging of the brain immensely. Still, while the images produced by this technique were much more detailed than with any other technique ever used, they were still snapshots of the brain and reflected structure more than anything else. Later in the 1970s, though, the PET (positron emission tomography) approach was developed. The idea with this technique is that if a nerve cell is active, blood should be flowing around it because of the need of oxygen for the nerve cell to work. That blood flow is detectable by an increase in positron emission that is then detectable by a PET machine. Measuring this phenomenon revealed some very interesting and precise information on brain activity during specific activities. The only snag is that in order to make such measurements the subject has to be injected with radioactive materials that can tag the blood flow and activity. The obvious danger of injecting radioactive materials into the body, coupled with both the relatively long times needed to do the scans and their limited resolution, has tended to push this technique aside.

The most widely used approach to mapping the brain has turned out to be a technique called fMRI. [Figure 101] This approach uses Magnetic Resonance Imaging (MRI, yes, that same technique they used to scan the knee you twisted at the Thanksgiving football game). The f stands for "functional," which means that it uses a proxy for activity—which is, again, blood flow. In comparison with the other approaches mentioned above, especially CAT and PET scans, fMRI has many fewer problems. In fact, fMRI can scan a single cross-section of the brain in less than two seconds, and can complete an enhanced full scan of the brain in less than two minutes. The resolution of the technique, that is, its ability to hone in on specific areas of the brain, is about two to three times as good as PET. A typical experiment using fMRI involves exposing a subject to a stimulus or outside challenge, as an MRI scan is made that images many progressive slices of the brain. A computer collects the images from the slices and reconstructs the picture of blood flow, hence of brain activity. The view of the brain we get from using this method is both exquisite and, in some cases, controversial.

David Dobbs, a science writer, has summarized the breadth of such experiments in the following passage:

Thousands of fMRI studies have explored a wide range of differences in brain activation: adolescents versus adults, schizophrenic and normal minds, the empathetic and the impassive. Researchers have used fMRI to draw bold conclusions about face and word recognition, working memory and false memories, people anticipating pain, mothers recognizing their children, citizens pondering ethical dilemmas—not to mention why many consumers buy Coke even though they really prefer the taste of Pepsi.

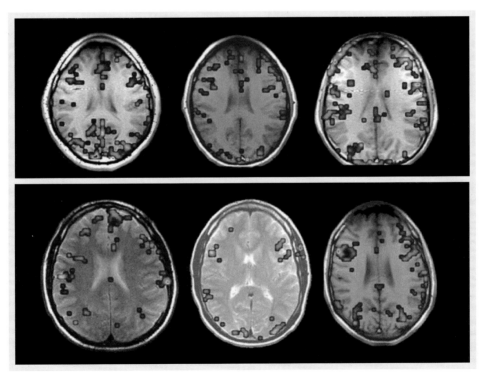

FIGURE 101. Functional magnetic resonation imaging (fMRI) pictures of regions of human brain activity during several different activities. Courtesy David Dodds.

Why is this approach controversial? While the idea that blood flow would reflect brain activity is a good one, some scientists doubt that it is the *best* indicator of brain activity, prompting critics of the approach to suggest that the pictures given by fMRI are often times over-interpreted. Perhaps even more difficult to quantify are problems with the design of fMRI experiments. The controls and the approaches to an fMRI scan like the one just described must be carefully chosen and implemented. Otherwise experimenter bias can creep into the collection of data and, even more harmfully, into the interpretation of those data.

Head Cases

Paleontologists have a knack for thinking of neat ways to bring their fossils to life. One method that is relevant to our discussion of the brain relates to the ability to estimate from a fossil skull how much brain would fit into it. This has traditionally involved seeing how much mustard seed it takes to fill a fossil skull, but estimating the volume of the brain that resided in a partial skull has always required some fancy calculation. The latest wrinkle is to complete a "virtual reconstruction," using CAT scans of what is left of the braincase, and then to derive the volume electronically. Whatever we do, though, when we look at the amount of brain that can fit into living and fossil skulls we see the following:

The average brain size for a modern ape species runs from about 350 cubic centimeters to about 500 cc. Known australopith brain sizes are in the 320 cc to 502 cc range (hardly any bigger relative to body size), though a small jump may have occurred in early *Homo*, with specimens often identified as *Homo habilis* coming in at around 510 cc to 750 cc. *Homo ergaster* brains are a little bigger, at about 800 cc to 900 cc, while what

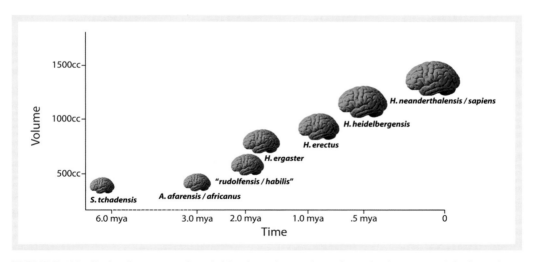

FIGURE 102. Brain size among hominids plotted over time. Over the long stretch before about 2 million years ago, hominid cranial capacities were comparable to those of living apes. A million years later brain sizes had doubled, and now they have doubled again. Note that this does not necessarily imply a consistent rise in brain size in a steadily modifying hominid lineage over the past two million years. Brain size increased independently, for example, in the lineages leading to *Homo neanderthalensis* and *Homo sapiens*; and the overall pattern is plausibly attributed to the preferential survival of larger-brained species. Illustration by Gisselle Garcia.

is loosely defined as *Homo erectus* has brains in the 900 cc-1,200 cc range, and *Homo heidelbergensis* comes in at about 1,200 cc to 1,300 cc. Neanderthals, with a huge documented range between about 1,200 cc and 1,740 cc, are in our own league, and indeed the upper figure, for a late Neanderthal from Amud, in Israel, is the largest fossil human brain ever recorded. [Figure 102]

At first glance, these figures make it appear that nothing much at all occurred in the brain-size department in the first 4 million or 5 million years of hominid evolution, while over the last 2 million years or so hominid history has largely been one of steadily increasing brain size. The problem with this perception, though, is that, throughout, there have been multiple hominid species, which means that we are not necessarily dealing here with brain size increase due to inexorable natural selection within one lineage. In fact, there may have been a very mild tendency toward brain size increase with time among the australopiths, and there was clearly an accelerated tendency of this kind in *Homo* over the past 2 million years; but this trend was just as plausibly due to the preferential survival of bigger-brained species as it was to within-lineage fine-tuning. Even less clear is the relationship between increasing brain size and increasing intelligence. Whales and elephants have some of the largest brains on the planet; and while they are clearly very complex creatures, any statement about their intelligence levels or their consciousness being equivalent to that of humans should be treated cautiously. Among modern humans themselves there does not seem to be any striking correlation between brain size and high cognitive skills. And while *Homo neanderthalensis* had a brain as large as ours, as we'll see in the next chapter this hominid species evidently did not possess our own hallmark of symbolic intelligence. Brain size *per se* is evidently not a perfect indicator of intelligence or cognitive complexity.

Brain *organization* is a separate matter, of course, though how the brain is structured internally is tough to read from its size and external contours—which are all, alas, that the fossil record offers. Nonetheless, the trend of increasing brain size in the evolution of the genus *Homo* does seem to point to a special role in this anatomical trait in our lineage, especially when we realize that the brain is a "metabolically expensive" tissue that does not come without its costs in energy consumption.

So what do we know about how the brain has changed in ways that make us human? In some ways a lot, but far from enough. As indicated earlier in this chapter, language, music, art and tools are critical in making us different from most other organisms on the planet. Because our perception of the world plays a critical role in these four typically human traits, perhaps it is something about our senses that differentiates us from other organisms. Fortunately, huge advances in understanding our senses have arisen from modern genomics. Explanations of how our senses work at the genomic level are important in two ways. First, we can show how the outside world is perceived and processed by the brain from a genetic and molecular perspective. Second, it is our senses and the evolutionary changes that occurred in their inner workings that make us so different from other organisms on the planet. The following are several examples of the tantalizing view we are getting from a genomic view of our brains.

Big Fish

Looking for differences in the genomes of humans and other organisms to pinpoint how we became human at this point is, we have to say, a fishing expedition. But, as any good fisherman knows, there are certain baits and tackle to use. A good nose for where to cast your line will also increase the probability of reeling in the big one. In the genomic case, the "big one" is finding a gene or suite of genes or a genetic process that has an impact on how language, art, music or toolmaking arose in our species.

There are two ways to catch the big one: good bait and tackle, and knowing where to cast the line. In our search for the genetic basis of the traits we have raised as important in becoming and being human, we suggest that our senses are a great first place to look. Because the whole genomes of humans, chimpanzees and a whole slew of other mammals have either been sequenced or are in the process of being sequenced, there are plenty places in the genome to cast our lines. Also, because the genes that code for certain aspects of our senses are often not just single isolated genes but rather clustered and repeated, this widens the areas of the genome where we can go "fishing".

With respect to better bait and tackle, a new genomic technique now exists that allows scientists to cast broadly about the genome for differences between humans and other organisms. This approach, using microarrays, has revolutionized not only our understanding of human evolution, but also medicine. Microarrays answer the tricky question of how we can query our genomes and the genomes of other creatures for differences in how genes make a human brain a human brain. After all, remember that humans and chimpanzees have very similar genes in their genomes. And, for that matter, the genes in the mouse and rat aren't that terribly different from human genes. A first step is to figure out *which* genes are active in human brains and *not* in other creatures' brains. What makes human nervous tissue human nervous tissue is the kinds of RNA (and hence of proteins) that the cells of the tissue make. So if we can take cells from the brain that have a specific function, and look at what RNA molecules are in those cells, we can infer which genes are important in making a cell behave a certain way.

Here is how scientists have overcome this huge problem. To understand this clearly, we need to return to all of the things we learned in Chapter 3. In particular, we need to remember the Central Dogma of Molecular Biology, which says that DNA (our genes) R RNA (intermediate messenger) R Protein. When DNA R to RNA we say that the DNA (or our genes) is being expressed, and call the RNA a gene product. It makes sense that if a particular gene is making a lot of RNA in a tissue like the brain, this gene is being expressed lot in that tissue. Now, if we compare the production of RNA in our brain and the brain of a chimpanzee, and we find that it is expressed at a high rate in our brain and not in the chimp's, we might really be on to something. The trick is, how can we do this?

If one is looking for a gene that might be involved in making our brain unique relative to, say, a chimpanzee's brain, a huge problem arises. Thousands of genes are being expressed in our brains, as well as in the brains of most other creatures. Add to this the fact that different parts of our brains are doing different things and hence might be expressing different genes, and the problem gets magnified even more. What with those thousands of genes being expressed and the multiple functions of our brains, we are not only faced with a fishing expedition, but also the problem of looking for a needle in a haystack.

Before whole genomes were sequenced, scientists could only ask if a particular tissue had a *single* gene that was producing RNA, and hence being expressed. The way scientists did this was to take the tissue, crush it up, and isolate the RNA from its cells. All of the thousands of different RNA molecules isolated in this way are a record of which genes are being expressed in that tissue. So now what? Well, the next step is, of course, to figure out which genes are expressed, right? But 20 years ago, scientists had very simple techniques that only allowed them to figure out whether a particular gene was being expressed. The way they did this was to take advantage of the fact that RNA is single-stranded and can form a double helix with another single-stranded nucleic acid, like DNA.

Now, if you take a particular gene you think is being expressed and is important in a tissue, and you have its DNA sequence, you can use this DNA sequence as a probe for the expression of the gene in the original mixture of expressed RNA. The double-stranded DNA from the gene we are interested in is denatured into single strands and labeled with radioactivity. Next, this labeled DNA is allowed to react with the original RNA mixture. When the radioactively labeled DNA probe finds its complement in the original mixture of RNA, it will form a double helix with its expressed RNA complement. Next, the mixture is "rinsed" so that anything that did not make a double helix gets washed out. And if it does make a double helix and is not washed out, it can be visualized by assaying for the presence of radioactivity in the mixture. If we don't see radioactivity, then the gene we are interested in is not turned on. If we do detect radioactivity, we can infer that the gene we probed our originally expressed RNA with is being expressed in that tissue. Whew! Now, imagine doing this experiment for 30,000 genes one at a time. Big job, lots of time, and lots of radioactivity.

But modern genomics has come to the rescue for studies where the expression of genes in certain tissues is very complex. The major technological advance in examining all of the expressed genes in a tissue was to miniaturize the whole visualization process and to find a technique that allowed scientists to examine all 30,000 genes simultaneously. And the tool that was developed for this is called the microarray – micro for miniaturized, and array for the way the genes of a genome are arranged in the visualization process.

In a microarray, the whole process of visualization is reversed compared with the old-school way of looking at expressed genes that we described. In a microarray, single-stranded DNAs from as many genes as possible are affixed in a rectangular array on a glass slide. The position of each gene's DNA is recorded. The RNA from the tissue being studied is isolated, and then every RNA molecule from every gene that was expressed in that tissue is labeled with a fluorescent dye (the use of radioactivity has fallen on hard times lately, and fluorescent dyes have been the reason) and allowed to react with the single-stranded DNA on the glass slide. Wherever a gene on the slide matches with an RNA molecule in the tissue, a small fluorescent dot will appear and stick to the array in the position where the gene was originally placed onto the slide.

In this way scientists can determine which genes are expressed and which genes are not expressed in different tissues. And they can do this for thousands of genes and several different kinds of tissues – *overnight!* With these genomic tools, we are set to go fishing.

Making Sense of Our Senses

Our five senses—and the balance between them—contribute enormously to the way we experience the world. Much as we may think we empathize with our pets, in fact humans (visual) and dogs (olfactory) live in completely different, if intersecting, worlds—even if those worlds would look identical to a robot. Lately, we have learned much about the molecular and genomic aspects of our senses, among the most important functions of our brains. Two of our five senses—taste and smell—more than likely have peripheral impact on language, art, music and toolmaking. This is not to say that some genomic-level changes that control these two senses may have had no impact on our becoming human. In fact, as we shall see, there are some very distinctive genomic changes in the genetics of these two senses that are distinctly human. But we *can* reason that the other three senses are more intricately involved in language, art, music and toolmaking. If we can pinpoint certain ways these senses are different in humans relative to other animals, we might be on to something. Let's look at some of the information that has recently come to light with respect to our senses.

TASTE AND SMELL

As mentioned above, these two senses are more than likely peripheral to an understanding of language, art, music and toolmaking. They seemingly are not important in how we see and process information for communication, but, as we will see, olfaction and taste can be very important for communication in other organisms. Recently, a Nobel Prize was awarded to olfactory researchers Richard Axel and Linda Buck for their pioneering work in figuring out some of the basics of olfaction in animals. Some of their results are intriguing, and indicate basic differences between humans and other animals. Such differences may be involved in our becoming human.

In humans there are more than 900 genes for olfactory receptors. These receptors are molecules made by genes in our genomes that interact with different odors. When a receptor interacts with an odor, a signal is sent to the brain indicating a positive interaction. When the mouse genome was compared with the human genome with respect to olfaction genes, it was discovered that mice have a larger number of olfactory receptors, up to 1,500. Humans and mice differ in their reliance on smelling things for identification of other individuals—and indeed it may seem remarkable that, as very visually oriented organisms, humans still have 900 or so genes for smell. Humans enjoy five basic tastes—sweet, salty, sour, bitter and umami (the novel tastes—to Westerners—that are elicited by some Asian foods). While scientists are only just beginning to understand the genomics of taste, some significant results have already been obtained. Bitterness, in particular, is a complex taste. Scientists have recently found 25 or so genes in the human genome that are involved in taste reception. Variants of these 25 genes in human populations have molded how humans perceive taste and may well be responsible for the various preferences for food in different human cultures. In fact, when all of the known variants of these bitter taste genes are compiled, humans can perceive bitterness in more than 150 ways. The other aspects of taste are not as well worked out as bitterness, but it is known that only three genes are involved in sweetness and umami perception. To date, the receptor genes for salty and sour taste have not been found.

The progress made in understanding the genomics of taste and smell is really astounding when one thinks of where scientists were even 10 years ago with respect to unraveling these senses. In one case – smell – we can show how we differ from other organisms like dogs and mice, and nicely explain the differences based on genomics. In the other – taste – we can demonstrate how variable humans are, and how the different versions of genes for this sense influence how people in different cultures might be reacting to different foods.

HEARING, TOUCH AND SIGHT

How our hearing apparatus develops, and the molecular and genomics basis of deafness, are both well-understood. Scientists even have some idea of how the human brain perceives music and language. Perfect pitch has often been suggested to be genetically controlled. In addition, a condition called Williams Syndrome, caused by the lack of two copies of a gene called elastin, may give rise to a range of effects among those with the syndrome. One of these effects is unusual musical ability, suggesting some genetic component for this trait. As with most complex traits, though, it is wise to interpret isolated cases with caution.

An important development from the completion of three full mammalian genomes—human, chimpanzee and mouse—has led to some very intriguing suggestions and hypotheses about the evolution of hearing in humans. Strong selection against certain alleles of genes important in forming the human hearing apparatus have led scientists to suggest that language development may also have involved how well we hear. An example is a protein called tectorin. Humans and chimpanzees have very different tectorin genes in their genomes, and not because of chance. It appears that strong selection has acted to mold the gene for tectorin in humans. Why is this important to language? Well, tectorin is a critical component of a membrane in our inner ear. Mutations in this protein cause the bearer to respond poorly to frequency changes in sound, and individuals with mutations cannot understand speech well. Andy Clark, the scientist who examined the chimp, mouse and human genomes in this study, suggests that the effect is like "replacing the soundboard of a Stradivarius violin with a piece of plywood." The suggestion that a single gene is important in language acquisition is not new (see the FOXP2 example below). The more candidates scientists come up with for this complex trait, the more quickly its mysteries will be unlocked.

FIGURE 103. If you see the number 42 written in a single shade, you might be color blind.

How the sensation of touch is transmitted to the brain is genomically the least-known aspect of all of our five senses. How we sense heat, however, is beginning to be understood from our genomes. Sensing heat involves a chemical reaction in our skin cells that sends a message to our brain. Six known genes are important as molecular thermometers. When the protein products from one of these genes (TRPV3) are stimulated by heat (greater than 91 degrees Fahrenheit), they become active, and open up gates on receptor-cell membranes that can then send electrical impulses to the brain. Presumably the same mechanism—but different TRPV genes—is responsible for responses to other temperatures. Other mammals have similar genes in their genomes for these molecular thermometers. How we sense spatial information with respect to touch is less well-understood, and this aspect of touch might be more relevant to our understanding of the human ability to make art and tools. However, given the advances made in our understanding of temperature sensing via touch it is not unreasonable to expect significant increases in our knowledge of this aspect of touch in the near future.

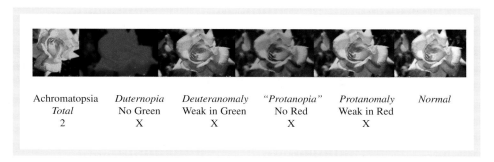

Achromatopsia Total 2	Duternopia No Green X	Deuteranomaly Weak in Green X	"Protanopia" No Red X	Protanomaly Weak in Red X	Normal

FIGURE 104. A rose is a rose. Most people will see the rose as in the normal picture. There are several kinds of color blindness that occur and these are listed below the pictures. How people with these color blindness syndromes would see the rose are depicted above. The chromosome where the color blindness gene resides is given below the name of the color blindness syndrome.

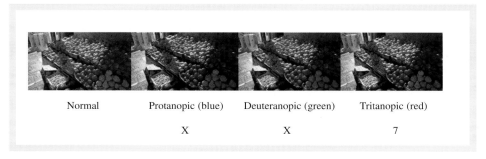

Normal	Protanopic (blue) X	Deuteranopic (green) X	Tritanopic (red) 7

FIGURE 105. How the fruit market would look if you were colorblind. The chromosome where the color blindness gene resides is given below the name of the color blindness syndrome.

Some of the genomics of sight are very nicely worked out, especially with respect to how the eye develops, detects color vision and orients visual signals. Color vision, which is most likely involved in how we perceive art, has been studied for decades, mostly because of the high incidence of certain kinds of colorblindness in humans. [Figure 115] Red-green colorblindness is found in about 8 percent of Northern European men. [Figure 116] This disorder is caused by genes called red and green opsins that are found on the X chromosome. Several different kinds of red-green colorblindness are caused by the opsin genes on the X chromosome—trichromacy, protanomaly, deuteranomaly, dichromacy, protanopia, and deuteranopia.

There is a third kind of color vision molecule called blue opsin, but colorblindness resulting from a lesion in this gene is rarely manifested. Why? One reason is that it is not found on the X chromosome. The total colorblindness that we talked about in the Pingalap Islands that resulted from genetic drift is called *achromatopsia* and has been traced to genetic lesions on Chromosome 2 and Chromosome 8. In this case, the eye cells that can detect different colors seem fine, but there is a lack of neurophysiological response to light. Scientists have recently narrowed the cause down to faulty genes that do not make the proper proteins for transmitting light responses via nerve cells to the brain.

Table of frequencies of colorblindness. Inc stands for incidence. X stands for X chromosome. 7 and 2 stand for chromosome 7 and chromosome 2 respectively. (http://webexhibits.org/causesofcolor/2C.html)

	Males	Females	Chromosome
Trichromacy	6.3	0.37	X
Protanomaly	1.3	0.02	X
Deuteranomaly	5	0.35	X
Tritanomaly	0.0001	0.0001	7
Dichromacy	2.4	0.03	X
Protanopia	1.3	0.02	X
Deuteranopia	1.2	0.01	X
Tritanopia	0.001	0.03	7
Total	0.00001	0.00001	2

A quick look at the genetic lesions that produce the three kinds of X chromosome reveals that most red-green colorblindness is actually more complicated than just simple mutations in the genes for red and green opsins. This was first realized by Jeremy Nathans, a colorblind researcher at John Hopkins University, who used his own DNA from his germ line in his experiments (please don't ask!). The red and green opsin genes are found right next to each other on the X chromosome. The red opsin produces a protein that deciphers red light (light with wavelength of 560 nanometers causes the red opsin to change conformation and signal the brain). Similarly, the green opsin gene produces a protein that deciphers green light (light with wavelength of 530 nm causes the green opsin to change conformation and signal the brain). [Figure 106]

The opsin genes sometimes play a game of swap during sperm and egg formation, exchanging bits of themselves or making extra copies of themselves, or even eliminating themselves, from chromosomes in the sperm and eggs. The process that generates more than one copy of each opsin and sometimes eliminates one or the other is complicated, but involves misalignment of the genes when sperm and egg cells are formed. These misalignments lead to most of the lesions in the red-green opsin part of the X chromosome. Recently, geneticists have also suggested that women perceive colors more adeptly than men. Wait a minute! Men who have the requisite red and green opsin genes in their genomes should be seeing red and green just as well as women, right? Not so fast. Remember, guys, the red and green opsin genes that control how humans see red and green are on the X chromosome. And men are X-challenged, which means men have a single X chromosome. So it turns out that the two X chromosomes in women, and the double dose of red and green opsins from these two X chromosomes, may very well confer upon females the ability to see red and green colors more vividly. Why? The geneticists who did this work, Brian Verrelli at the State University of New York at Stony Brook, and Sarah Tishkoff at the University of Maryland, suggest that this phenomenon may be an old one that dates back to the hunting-gathering days of humans. While males were out hunting, females were gathering. The ability to see various shades of

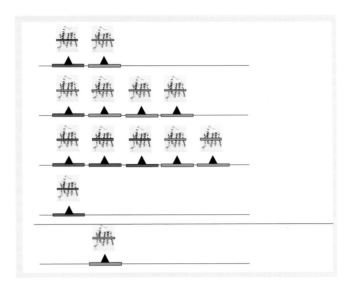

FIGURE 106. Representive arrangements of red and green opsins on the X chromosomes of humans. The top three arrangements result in normal color vision. Lesion 4 and 5 result in severe red-green colorblindness.

red was apparently more important for females as gatherers, in order to avoid the very nasty crimson red fruits that were poisonous and to take, instead, the softer red fruits that were edible. Maybe. [Figure 107]

In people who are not synesthetes, light enters the eye and the information from the light proceeds to the occipital lobe of the brain. Once there, the information is parsed out by the occipital lobe into its component attributes – color, form, motion and depth. Each of these component attributes is then sent to the areas of the brain where the information is processed. What happens in the case of the number/color switching synesthete is that number information is sent to an area of the brain that is near where color processing is located. Scientists think that this is where the information gets switched. Integration of genetic data for this syndrome (or rather, we should say, group of syndromes, because the different synesthetic effects almost certainly are different phenomena) would advance our understanding of how our senses are integrated with perception and consciousness.

While we have fished around a bit here by delving into our senses of sight, sound and touch, it should be obvious that the genomics of these senses is a good place to start if we want to understand what makes us unique. Another way to figure out where the best "fishin' hole" is for this endeavor concerns focusing on specific human aptitudes such as language or music. This kind of fishing expedition has been fruitful, too.

A Gene for Language?

Seven language disorders are commonly trotted out when scientists consider gene candidates for language (see the table). All of these disorders have had molecular genetic explanations attached to them as a result of genetic and genomic studies. This relatively small number of genes is by no means exhaustive, yet the list does show the great promise that genomics has for determining one of the most important traits that makes us human—language.

One of the most interesting and most cited examples is the FOXP2 gene. In some humans a syndrome called developmental verbal dyspraxia, or DVD, causes an inability to move the mouth correctly in sequence. In addition, there are grammar deficits and repetitive language usage in individuals with DVD. While the disorder is complex, one form of DVD, called SPCH1, is caused by a simple Mendelian factor. FOXP2 has been shown to be the genetic locus involved in this syndrome. This gene makes a protein of the kind

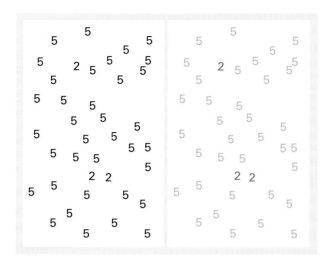

FIGURE 107. The strange cases of synethesia that exist in human populations have been cited as examples of the importance of the integration of color perception with consciousness. Synesthesia is a syndrome whereby those affected experience the blending of two or more senses. One of the more common versions of this syndrome is the switching of signals for spatial representation in the brain to colors.

Let's do a little experiment with perception. Type a clump of fifty or so 5's typed in black ink with two or three 2's mixed in. Now show this to anyone. Most people (estimates of different kinds of synesthesia range from 1 in 250,000 to 1 in 2000) will have some difficulty and take a little time to differentiate between the 2's and the 5's in such a clump. Now go back to your typing and color the 5's with green and the 2's with red—this is how synesthete who mixes spatial representation with colors would see the numbers. Such a person would be able to immediately pick out the 2's and 5's because the 2's and 5's will be represented by colors in their brains. This mixing of the signal for a spatial representation of numerical figures and colors is only one mix-up that synesthetes endure. Others include smelling colors, or tasting shapes. While no gene candidates exist for this syndrome, it is a good bet that synesthesia is controled by genetic components because it travels in families. By the way, people who are musically and artistically creative tend to be synesthetes at an unusually high rate. The syndrome is thought to be ultimately controlled by a swapping of signals in the brain.

known as a transcription factor. Transcription factors most commonly regulate the development of organisms. In humans, FOXP2 happens to regulate how brain circuitry forms in several very different brain regions. FOXP2 is active in the cortex, striatum, thalamus and cerebellum, and is obviously a very important protein in brain function and development. While the exact details of its involvement in language have not yet been determined it has been shown, using the same approach that Andy Clark used on the tectorin gene, that while FOXP2 is highly conserved in vertebrates, the gene underwent considerable acceleration of change in the human lineage after humans split from chimpanzees. In other words, there has apparently been intense natural selection for these changes.

While not entirely pinning down the genetic basis of language, the FOXP2 story is an interesting and encouraging one with respect to our understanding of language origins. In our opinion, though, the possibility that this approach will lead to recognizing genes "for" language, art, music or tool making is remote. These human attributes are extremely complex, and single genes almost certainly do not control these important and solely human features. While the gene-by-gene approach has given scientists a lot of information about language, music, art and toolmaking, it is the microarray approach that promises to open the door to an understanding of how the many genes involved in these attributes interact to result in our uniquely human abilities.

SYNDROME	TARGET	EFFECT	GENETICS	EVOLUTION
primary microcephaly, MCPH	brain	head circumference small; moderate mental retardation	Six genes; neurogenesis	higher selection on these genes in humans vs. chimpanzees
Joubert Syndrome, JBTS	brain	brain stem malformed; developmental delay	one gene; neural axon formation	higher selection on this gene in humans vs. chimpanzees
BFFP; bilateral frontoparietal polymicrogyria	brain	abnormal folding of layers in brain; mental retardation	one gene Involved in cortical patterning	not studied in apes
specific language Impairment, SLI	language	verbal and nonverbal skills are unmatched	large chromosomal region; involved in synapses	increased number of these genes might be important in humans
developmental dyslexia, DD	language	reading and spelling difficulties; impaired language processing	at least six genes	some of the localized genes have different sequence signatures in chimpanzees
autistic disorder, AD	language	deficit in reciprocal social interaction and communication	complex; many chromosomal regions	no direct association with language yet
developmental verbal dyspraxia, DVD	language	impaired learning and ability to move mouth in Sequence	one gene involved; FOXP2	FOXP2 has undergone rapid sequence changes in the human lineage

Dissecting the Primate Brain

As genomics has developed, the microarray technology discussed earlier in this book has become more and more important in detailing the similarities and differences in the genetics and development of organisms. This is because array technology can allow you to do experiments that were unthinkable even 10 years ago on what genes are turned on in tissues. If you want to know what genes are actively different in a chimpanzee and a human, one way to do this would be to ask what the differences are, gene by gene. Because thousands of genes are involved in making and maintaining a brain, this task would be thousands of experiments long and would very tedious, indeed. But if, instead, you ask the GENOME what genes are turned on an off in one experiment, the task becomes one that is eminently doable.

This approach is exactly what several groups have tried in understanding the differences between chimpanzee and human brains. The procedure is a relatively simple one, and the results are stunning if only for the sheer magnitude of information obtained. While some scientists emphasize the problems that exist in these kinds of studies, the promise of understanding the contrast of gene expression in the brain of a chimp and the brain of a human is nonetheless amazing. The chimpanzee/human experiment is accomplished like any other expression study using microarrays. The RNA from a particular tissue is obtained and labeled with a molecular marker, usually a fluorescent dye. Because the brain is such a complex mass of tissue, scientists who do these studies work with specific parts of the brain. These specific parts are regions of the brain that have been previously suggested as important in selective human brain activity.

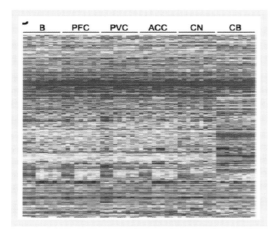

FIGURE 108. Microarray picture showing the levels of gene activity in the different parts of the brain shown in the previous figure. Courtesy of Svante Paäbo.

The RNA representing turned-on genes in the particular tissues is labeled with different "colored" fluorescences in the human tissue and the chimp tissue. Then the two kinds of labeled nucleic acids are hybridized to an array of thousands of human genes. When the "smoke clears" in the hybridization step, the picture of the microarray tells the observer which genes are differentially turned on or off in chimp and human tissues. A typical result of this kind of experiment is shown in Figure 108. In this figure the human genes that are turned on glow red, and the chimpanzee genes that are turned on glow blue. Tissues are listed across the top, and genes are thin slivers running in rows across the array. Whenever a gene makes a gene product in both chimpanzee and human, the slide is white. Only genes that are expressed in either or both chimp and human are used to make this array. As you can see, there are some distinctive differences between the expression in the brain of human and chimpanzee genes over all of the various kinds of tissues represented in the experiment.

The sheer quantity of information in these kinds of studies is somewhat overwhelming, but these results and the development of technology to analyze them hold great promise for understanding the genetic basis of differences in brain function between us and our close living relative, the chimpanzee. Several microarray studies that have examined which genes are being expressed and at what level have come to some surprising conclusions. What has been discovered so far is that there are specific kinds of genes that are exclusively expressed in human brain tissues. One set of microarray experiments used cerebral cortexes of humans, chimps and macaque monkeys. The results of this study indicated a large number of genes (nearly 200) in humans that are expressed at a much higher rate than in chimpanzees and macaques. The higher levels of gene expression in humans are referred to as "up-regulation."

Because there is some worry that different parts of the brain are doing different things another research group, under the direction of Svante Pääbo at the Max Planck Institute in Leipzig, Germany, took five regions of the brains of humans, chimps and macaques and examined the tissues from these five different regions (see Figure discussed above) separately for the three species. This group found that the various regions of the brain—especially the cerebral cortex, the caudate nucleus and the cerebellum—have very different expression patterns when humans and chimps are compared. Furthermore, this study showed that nearly 10 percent of the genes in the five brain regions differed between humans and chimps. Again up-regulation in humans appears to be the mode by which human and chimp brains differ.

These results might not be that surprising, though. In 1975, Marie Claire King and Alan Wilson at the University of California, Berkeley made the prescient suggestion that chimps and humans matched at 99 percent of their genomes. In other words, while we obviously look and behave very differently from chimps, there is only a 1 percent difference between your genes and those of a chimpanzee. King and Wilson then suggested that the reason we look so different from chimps lies in the *regulation* of the genes. If a single gene in a human controls how 50 proteins are expressed (this happens in gene regulation in humans and most other organisms), then by altering a single gene, a large effect in expression can be implemented.

A lot of work has yet to be done, but these early array studies comparing how genes are expressed differently in chimps and humans are intriguing. The next obvious step is to figure out whether there is a pattern to the kind of genes that are being up-regulated. Also, because it appears that these up-regulated genes are not changing in their sequences and rather are being up-regulated by other proteins, the big job will be to figure out what the regulators are. It would not be surprising if scientists learned in the quite near future which of the genes in the genome are expressed in the brain, and how they all interact.

Wrapping Your Brain Around It – A Summary

Human beings are different from their extinct and living relatives in many ways. For example, only we walk on two legs, and that has made an enormous difference to what we look like. But to us, the difference that really matters is that we are capable of thinking abstract thoughts, and of talking about them. Most importantly, it is our intelligence that makes us different. This intelligence is the product of our brains which, we hope we have been able to show, we are now beginning to understand far better than ever before from the molecular, genetic, developmental and functional points of view. Even so, we are still a very long way from being able to see just how the genetics, chemistry and architecture of our brains translates into the consciousness we experience.

Can a historical perspective help? We can see from the archaeological record that, over time, our precursors made more and more complex stone tools. And the fossils show us that over the past couple of million years the brains of our extinct relatives increased remarkably in size. The obvious message here is that they were getting smarter, although not necessarily slowly and steadily. Intelligence seems to have increased in a series of steps, in a pattern we still don't fully perceive. But still, it doesn't seem that this increase in intelligence was just like climbing a ladder. As far as we can tell, the *kind* of intelligence we have is different from that of any other living animal species, including some of our extinct relatives. None of them made paintings or sculptures. And most likely, smart though they were, they didn't tell each other stories, either.

One thing that is clear from looking at the human fossil and archaeological records is that, although both proceeded in a series of jumps, those jumps did not happen in synch. You can't use the appearance of a new hominid species to "explain" a new behavior, which actually makes sense, because any inventor of a new way of doing things must belong to a species that already exists! We can be sure that our ancestors at one time did not have abstract thought or language. Yet today, we have both. How did the change come about? Well, the earliest *Homo sapiens* seem to have behaved pretty much like earlier hominids that lacked these features, even though they were physically very different. And, because we can fairly assume that the brain wiring that permits complex thought was acquired along with all the other striking physical features of our species, it must have been a *cultural* rather than a *physical* innovation that sparked abstract thought – in some population of the species *Homo sapiens*. What might that innovation have been? This is the subject of the next chapter.

CHAPTER NINE

THE IMPORTANCE OF LANGUAGE

N O MATTER HOW WE TRY TO EMPHASIZE THE MANY OTHER WAYS WE DIFFER AS *HOMO SAPIENS* FROM EVEN OUR CLOSEST REATIVES, FOSSILS AS WELL AS

living, our focus inevitably returns to our cognitive differences from them: the unique ways we perceive and process information about the world around us. To the best of our knowledge, we human beings are most profoundly set off from the rest of Nature in being symbolic creatures. Other animals walk on two legs, or have relatively large brains compared with their bodies, have complex vocalizations, or even use tools; but only we *think* as we do. That is to say, we human beings mentally represent the material world, and our own experiences, by discrete symbols – and then remake the world in our minds by combining and recombining those intangible symbols. As far as we know, no other animals do this, or indeed ever did.

Other organisms react more directly to the stimuli coming in from their environment, and often they do this in very sophisticated and subtle ways. But none seems to reconstruct the world in its head as we do, so that all experience their physical surroundings and social circumstances pretty much as Nature presents them. Of course, perceiving the world in this way does not obviate reasoning or understanding of some kind, for intuitive processes can clearly govern very complex reactions indeed. What's more, even our own responses today are, to a high degree, still governed by unconscious reasoning, the bequest of a very long history of brain evolution. Inside our skulls we still have "fish brains," "reptilian brains," and "primitive mammal" brains, not to mention all the neural apparatus that apes have at their disposal. But it appears that only we have this overlayering of symbolic mental manipulation, this ability that allows us to exploit the world around us in entirely unprecedented ways. We don't do what our predecessors did, only a little better. Our unique behaviors represent a qualitative leap.

We are, then, separated from the rest of Nature by a profound cognitive chasm. Yet it was not that long ago, it seems, that our immediate ancestors were non-symbolic animals themselves. How and when, then, did our precursors bridge this chasm? Interestingly, it was this issue that caused the only major intellectual breach that ever fissured the relationship between Charles Darwin and Alfred Russel Wallace, the co-inventors of the idea of evolution by natural selection. Darwin, intent on finding natural explanations for everything he observed, was content to ascribe modern cognition to the outcome of eons of natural selection, forever seeking out more favorable variants to favor. But the action of natural selection implies continuity, and Wallace – although enthusiastically selectionist on all other matters – was disturbed by the *dis*continuous nature of the cognitive difference between us and other organisms. [Figure 109] He simply couldn't

FIGURE 109. Alfred Russel Wallace, seen here in a late portrait taken at his home in southern England, was the co-inventor with Charles Darwin of the theory of evolution by natural selection. Nevertheless, he had legitimate doubts about how natural selection could have propelled modern symbolic consciousness into existence.

see how natural selection could arrive at a result like this. Lacking any alternative explanation Wallace, who must have been one of the most decent and intellectually generous characters who ever lived, turned to spiritualism, highly fashionable in the London of the 19th century. Blinded by an utter guilelessness of character, he was simply incapable of detecting fraudulence in the charlatans who profited from the spiritualist craze.

History has pilloried poor Wallace for his naïveté ever since; but in fact, his basic perception was an accurate one. Natural selection is not a creative force; it cannot call the new into existence in a relentless drive toward optimization. Instead, as we've already pointed out, it can only work with what is there already. Of course, Darwin had a good point, too. Our unique cognition is the product of our brains; and our brains, as we've seen, have an extremely long and accretionary evolutionary history: a history in which natural selection, paramount in the Darwinian view of things, must have played a role alongside all the other factors that promote change over the evolutionary history of any group of organisms.

A hundred years and more later, we now have a pretty good fossil record of hominid brain evolution to consult, not to mention huge advances in our knowledge of how the brain works. As a result, we are privileged to have a much fuller perspective on this matter than was available to either Darwin or Wallace. Still, even this new knowledge is not particularly easy to interpret. For a start, even though we now know a whole lot about the fine structure of the brain, and which bits of it are active in different situations, we still really have no idea how a mass of electrical and chemical signals in this very complexly wired structure actually gives rise to what we experience as consciousness and everything that goes with it. On the other side things are not that much better, because problems in interpreting the fossil record are legion. The most important of these problems stem from the difficulty we introduced earlier of identifying the actors in the human evolutionary play. We know that, after a long period when hominid brain sizes appear to have remained pretty stable, around 2 million years ago they began to increase dramatically in size. Starting off at around a third the volume of our own, by a million years ago they averaged up to about half, and then by about 200,000 years ago some hominids at least (and not just *Homo sapiens*) had brains as big as ours today – which are seven times or more larger than what we'd expect to find in an average mammal of our size. This is a dramatic increase, and something important must have been going on to promote this result.

The problem is that we don't know what that something was. And we won't really be able to make any educated guesses until we understand exactly what was happening in this trend toward increased brain size. We know by now that this increase was not a straightforward matter of generation-by-generation enlargement under the beneficent hand of natural selection. Not only have environments fluctuated far too rapidly over this period for consistent directional selection to occur, but the emerging picture of human evolution is one of species diversity and of an ongoing exploration of the different ecological opportunities available to hominids. If we are thus not simply following a single chain of brain enlargement through time, the picture suddenly becomes more complex. Instead of individuals within a lineage steadily gaining brain mass from generation to generation—which admittedly seems the commonsensical view to members of an intelligent species, who find it self-evident that even a little bit more intelligence would be a big evolutionary advantage—it seems plausible that within a diversity of hominids, it was the bigger-brained species that had the edge: that those species with larger brains, and thus presumably with greater "intelligence," out-competed the less brainy ones.

At the risk of belaboring the point, accurately defining the species involved is extremely difficult at our present state of knowledge. Moreover, even if we knew with any precision what those species were, what the ranges were in the size of their brains, and how long they resided on Earth, it would still be a problem to know exactly what it was about the successful and larger-brained ones that made them win out. Still, we know that the trend toward larger brain size is a real one. Hominid brains really did get bigger, on average, over the past 2 million years. Not only that, they seem to have gotten bigger in multiple hominid lineages. Even though the record is slender, there's no doubt that brain volumes increased in the lineage leading to our big-brained selves. We also have a pretty good record of brain size increase in the separate lineage that culminated in the big-brained Neanderthals. And even in the far-flung hominid outpost of eastern Asia, late populations of *Homo erectus* (which eventually may be recognized as a separate species) had bigger brains than their earlier counterparts.

These parallel trends toward greater brain size went on for a very long time, during which the old pattern of multiple hominid species in the Old World at any one time continued undisturbed. Yet, as we'll see in more detail, when modern symbolic *Homo sapiens* came along, the competition suddenly disappeared, everywhere. And it was evidently not the big brain of *Homo sapiens* that by itself made the competitive difference; it was symbolic thought, as substantiated by archaeological evidence of symbolic behaviors. It was this that changed the rules of the evolutionary game. In any event, whatever exactly it was that promoted brain size increase in multiple hominid lineages over the past 2 million years, it was clearly not the increasingly complex manipulation of mental symbols over the generations. Symbolic thought appears to have been an entirely new and effectively unanticipated development. Whatever it was that underwrote the prior hominid brain size increase, it didn't have anything to do with increasingly sophisticated symbolism. We wish we knew just what it was.

The Origin of Symbolic Thought

We looked briefly earlier at the emergence of *Homo sapiens* as an anatomical entity. And we noted that, whereas the first anatomically modern *Homo sapiens* in Israel were able to coexist in some way with *Homo neanderthalensis* for an extended period, the Neanderthals disappeared quite promptly after both the incursion of the behaviorally modern Cro-Magnons into Europe, and the development of a Cro-Magnon-equivalent stone tool technology in Israel. Quite evidently, in *Homo sapiens* the origin of "modern" behaviors lagged behind that of "modern" anatomy by several tens of millennia, at least. This hardly comes as a surprise, given the disconnect between anatomical and behavioral innovation that we see throughout hominid history; but because it involves the bridging of the cognitive chasm we spoke of earlier, it probably demands explanation more loudly than any other event in hominid history does.

Before tackling this matter head-on, though, it is worth looking at the evidence we have for the origin of our unique modern thought processes. Thought itself is, of course, an intangible, leaving no direct traces. So while, as we've already seen, the archaeological record gives us only the most incomplete perspective on the evolution of technologies, the situation is very much worse when we approach the area of cognition. Stone tools are only a small part of the complete technological panoply of any species, but at least they are a tangible indicator of something that was going on at any point in time, for stone tools are preserved remarkably well. On the other hand, thoughts and perceptions aren't, or at least they weren't until the invention of writing, a mere 5,000 years ago. We thus have to read mental processes from the most indirect evidence, and even then we can be sure of nothing.

We can, of course, reasonably infer from the occasional leap in the conceptual complexity of stone tools as time passed that hominids were becoming more "intelligent." But what such intelligence meant in concrete terms is impossible to say. Even today, "intelligence" expresses itself in myriad ways; and nobody, however smart, is good at everything. The way the world is actually perceived through the lens of a particular type of intelligence is yet more elusive, for it is mediated by a host of cultural influences. Still, no difference could be more striking than that between the material records left us by the aboriginal inhabitants of Europe, the Neanderthals, and by the Cro-Magnons who came in to replace them beginning around 40,000 years ago. Here, for the first time, we have clear evidence for the appearance of a *qualitatively different* kind of intelligence.

For earlier times, we can be reasonably confident in speaking vaguely of an aggregate "increase" in hominid intelligence over long spans of time (though for the most part it was business as usual). But the Cro-Magnons were not just smarter overall than the Neanderthals, whatever we might mean by that. They were *differently* intelligent. As gifted and intuitive as the Neanderthals doubtless were, they can fairly be described as the most sophisticated development yet of what had gone before. The Cro-Magnons were all that, no doubt. But they were something in addition. They were *symbolic*, and we don't need to infer this from indirect indicators such as their really cleverly made stone tools, or the neat implements they made from other materials, such as bone and antler, showing exquisite sensitivity to the mechanical properties of those particular materials. They bequeathed us items that were *overtly* symbolic: clear proclamations of symbolism and symbolic thought processes.

Well before 30,000 years ago, Cro-Magnons were creating astonishing art on the walls of caves. The best early example of this is the astonishing animal art of the limestone cave of Chauvet, in southern France, some of which is dated to 33,000 years ago. Experts had guessed it, from its sophistication, to be half that age—until new technology

FIGURE 110. Abstract symbols composed of dots, lines and shading in red ochre, from the cave of El Castillo, in Spain. Although the decorated Ice Age caves of France and Spain are most famous for their vivid animal images, they also abound in symbols of this kind, many of which are repeated at multiple sites. Copyright © J. Herrer and Asociación Cántabra para la Defensa del Patrimonio Subterráneo.

permitted radiocarbon dates to be made on the charcoal used to execute the drawings. In these deft representations of horses, woolly rhinoceroses, lions and a host of other animals, we see some of the earliest known expressions of what we intuitively and forcefully recognize as art—even though a consensus definition of what "art" is eludes us. And although this was early art it was not, in any sense of the word, "crude" art.

We do not see a learning curve in what is often known as Ice Age art (these paintings and drawings were made as the last Ice Age in Europe was closing in on its most severe stage, before gradually warming from about 18,000 years ago). Instead, some of the earliest art (as today, some is pretty indifferent) is among the most powerful and most meticulously observed of any ever created. From the very beginning, the cave art of the Cro-Magnons reflects an entirely mature, "modern" sensibility.

What immediately attracts our attention on the walls of the Ice Age caves is, of course, the animal representations. These are clearly symbolic; after all, what could be a more overt symbol than a graceful image of a deer drawn 20,000 years ago on the wall of a cave in Spain? [Figure 110] Surely, there is a one-to-one correspondence between the real deer on the landscape and its representation in the cave. This is true, as far as it goes. But there was evidently more to this art than mere representation. The animal images are often accompanied by strange geometric symbols, meaningless to us today but almost certainly pregnant with meaning to the artists who made them. Indeed, it may well be that in a more abstract sense the animal images were equivalent to the geometric signs, each with its own meaning, hidden to the uninitiated such as ourselves. Almost certainly all of these images, representational and geometric alike, were part of a wider symbolic system: the reflection of a system of beliefs that encompassed the ways the artists saw the world around them and explained their place in it. And what could be more human than that?

The early Cro-Magnons didn't stop at cave art in the production of symbolic pieces. Probably even older than the Chauvet paintings are the world's earliest known art works, from cave sites along the side of a river valley in southern Germany. There, the cave of Vogelherd has yielded a series of exquisite small animal figurines, among them a horse, a mammoth, and a lion. The horse, polished from years of being worn as a pendant around someone's neck, is particularly remarkable because it is no literal rendition of the chunky, rather pony-like horses that roamed the steppes of Ice Age Europe. [Figure 111] Instead, with the flowing lines of its back and belly and the graceful arching of its long neck, it is a perfect evocation of the abstract essence of all horses: symbolic in every sense of the term. Symbolic in another way is a 10-inch figure from the nearby site of Hohlenstein-Stadel. This strange piece has the body of a man (or maybe a woman) but the head of a lion: a classic example of the kind of reordering and recombination of symbols that goes on in the human mind. The probability that it represents a mythical character, part of the larger system of beliefs of a long-vanished society, is enhanced by the recent discovery of a similar figure in another cave in the same valley.

About as old are some flat plaques of bone from a rock shelter in southwestern France that bear lines of incisions and punctuations made with an engraving tool, without question evidence of a system of notation. It has even been suggested that one of these represents a lunar calendar. Whether or not this is correct, again we see dramatic evidence for symbolic mental processes and activities. It is in this early period too, well over 30,000 years ago, that we find the first evidence of music, in the form of flutes made

FIGURE 111. Polished by years of wearing as a pendant, this exquisite tiny scupture of a horse was found at the cave of Vogelherd, in southern Germany. At well over 30 thousand years old, it has a good claim to be the world's earliest known work of art. Photograph by Alexander Marshack.

from hollow bones, with multiple holes and complex sound capabilities. We'll never know exactly what the music that was played around those Upper Paleolithic campfires sounded like, but modern musicians have managed to extract haunting melodies from replicas of such instruments.

If those early Cro-Magnons made music of this kind, there can be no doubt that they sang and danced as well. But just as suggestive as the existence of these early musical instruments is what they are made of. Thirteen of the 14 now known were made from the same bone from the same species—part of the wing of a large vulture. Bones of these creatures are simply not otherwise found at Paleolithic living sites, which are generally littered with animal bones, the remains of ancient meals. Clearly, this particular element was sought out, possibly just for its physical qualities, but equally likely because of some symbolic association that existed only in the Cro-Magnon mind.

The Earliest *Homo sapiens*

We could continue in this vein, but a complete account of Ice Age art and technology would fill several volumes, and we think that by now the point has been made: The Cro-Magnons were *us* in every significant sense of the word, biological and behavioral alike. Were a time-machine to transport us back to Cro-Magnon times, we would doubtless feel culturally uncomfortable, just as anyone does who encounters a new and vastly different culture for the first time. But we would clearly feel ourselves to be among fellow human beings, with all of the same basic capacities that we have today. When we go back further in time, though, the picture becomes rather blurrier. As we have mentioned, the earliest intimations that we have of symbolic activities are found in Africa. The most famous of these come from the Middle Stone Age cave of Blombos, near Africa's southern tip.

Not long ago, in a level dated to about 75,000 years ago, two small plaques of ochre were discovered at Blombos that bore simple incised geometric markings. [Figure 112] There is ongoing argument about the significance of these markings, but many believe that these are symbolic engravings, and hence evidence—albeit far less impressive than what we see much later in Europe—of symbolic thought processes. Around the same age, also from Blombos, are some small invertebrate shells bearing holes. The holes have been described as piercings, to allow stringing of the shells as beads in a necklace, although there are alternative explanations of how they might have gotten there. Taken together, however, these findings suggest that symbolic thought processes may at least have begun to stir among hominids living near the tip of Africa at that early time. Unfortunately, of the many other Middle Stone Age sites known in the area, none has yet yielded any comparable objects.

Some way farther to the east along the southern African coast, though, lie the sites of Klasies River Mouth, with evidence of sporadic occupation by hominids between about 115,000 and 60,000 years ago. [Figure 113] Here, the South African archaeologist Hilary Deacon infers from the apparent use of their living space by the hominids that this space was divided up in a manner that can be described as symbolic. As we've already mentioned in passing, Deacon has also reported convincing evidence of cannibalism at Klasies River Mouth, but whether this behavior can be regarded as uniquely modern is doubtful; it has been quite plausibly inferred from those 800,000-year-old hominid bones found in Spain.

FIGURE 112. Not far from Africa's southern tip, Blombos Cave was inhabited by humans around 75 thousand years ago. This ochre plaque bears an incised geometric design that may qualify it as the world's earliest known symbolic object. Courtesy of Christopher Henshilwood.

The American archaeologists Sally McBrearty, at the University of Connecticut, and Alison Brooks, at George Washington University, have extended Deacon's cultural interpretations to even earlier sites in Africa, noting that certain features only associated with modern humans much later in Europe are found in Africa at much earlier times. Among these are not only sophisticated harpoon points from the eastern Democratic Republic of Congo that may be as much as 60,000 to 80,000 years old, but also such features as ochre use, blade manufacture, flint mining and long-distance trade in valuable materials. On this basis, McBrearty and Brooks have argued that the "Human Revolution" seen in Europe with the arrival of the Cro-Magnons was a purely local event, and that in Africa, in contrast, a slow awakening of modern human consciousness is seen over the past 250,000 years or so. Still, it is arguable whether we can read symbolic thought (as distinguished from advanced intuitive processes) from any aspect of the stone tool record, and the few cases of claimed early symbolism in Africa are all debated. It has also been contended that much of the evidence adduced for early African symbolism is, in fact, no more impressive than some of the evidence left behind by Neanderthals in Europe.

What nobody plausibly disputes is that the anatomical entity *Homo sapiens* arose in Africa, as those very ancient specimens from Herto and Kibish in Ethiopia suggest. But to confuse the issue, no hominid bones are known from Blombos, and at Klasies River Mouth the hominid fossils – the cannibalism victims – are very fragmentary. [Figure 113] More strikingly, however, most of the Klasies hominids do not appear to be typical anatomical *Homo sapiens*. Obviously, there is a lot to be learned about what was going on in Africa in the crucial periods spanning the emergence of *Homo sapiens* as both an anatomical and a behavioral being. It is nonetheless clear that *Homo sapiens* arose as an anatomical entity well before it emerged as a behavioral one, and that the first intimations of "modern" behavior patterns come much later. If what we see at Blombos and Klasies is indeed as claimed, then what we are witnessing there are the initial glimmerings of a capacity whose full potential had to be "discovered" over tens of millennia – if, indeed, it has yet been fully explored.

Where Language Enters the Picture

Homo sapiens shows very significant differences throughout its skull and its body skeleton from any of the "archaic" hominids we know. Whatever the nature of the genetic event behind the biological reorganization marked by the appearance of our species as a physically recognizable entity, it resonated widely. It seems entirely logical to conclude that one of its results, invisible as it is to us today, led—when added to everything that had gone before—to a brain that was capable of producing symbolic thought. Yet evidently, that brain continued to function in the old way for a considerable period of time, until some stimulus set it on to a new course. Because the biological innovation was already in place, it seems that this stimulus must have been a cultural—or at least a behavioral—one. What might it have been? We could ask the old question, "What is it about our behavior that makes us human?" and trawl though the possibilities with this question in mind. Unfortunately, we'd wind up, as Don Brown did in the last chapter, with a huge long laundry list of characteristics, each of which would seem to have an equal claim on our attention. That would get us nowhere, for the essential human capacity seems to be a supremely generalized one that expresses itself in an unending array of different behavioral consequences.

Behavioral biologists have endlessly debated the matter of why we acquired this capacity. Nowadays it is fashionable to look toward social factors in the search for the driving factor of increasing intelligence, because our behaviors are mediated by our large brains, and human beings and their closest living relatives are so intensely social. A current favorite, which hangs on the complex behavioral interactions we see among individual human beings—and higher primates in general – is "Theory of Mind." This is a fancy term for the ability of one individual to read what is going on in another's head, and thus to be able to anticipate what he will do next. Obviously, this is a great advantage to any individual in a social group within which there is competition for resources and reproductive access; and it can exist at a number of levels of complexity, often grandly referred to as "orders of intentionality." If I believe something, that's the basic level. If believe that you believe something, we are already at a second order of intentionality. And if I believe that you believe that I believe something, we are at a third, and so on. Pretty soon the head begins to spin, and most people today max out at around five orders of intentionality, maybe six. It's obviously tough to gauge what's going on in a nonlinguistic primate's head, let alone to know how much he can guess about what's going on in another's; but a number of ingenious psychological tests have suggested that chimpanzees have at least a rudimentary Theory of Mind, at the lowest levels of intentionality, while baboons barely make it on to the scale if at all.

The suggestion is that from a chimpanzee-like start, the increase in hominid intelligence has been spurred by pressures toward increasing intentionality via a sort of feedback loop: The more cleverly you can anticipate what other individuals in your group are going to do, the more easily you can outmaneuver them for access to females or the more desirable males, which will increase your reproductive success, which will increase the numbers of offspring around who have your improved smarts, and on and on. The problem here, of course, is that if this feedback loop is such an effective mechanism for intelligence increase, why hasn't it happened in all lineages of social primates? Even chimpanzees have only just made it on to the intentionality scale after the passage of millions of years since we shared an ancestor, despite the fact that they live in complexly structured social groups, within which extraordinarily intricate relationships develop among individuals that involve politics and deception (to the extent that there's a difference), and all sorts of constantly shifting reciprocal interactions.

To operate effectively, this kind of mechanism—and this holds for nearly every behavioral stimulus you could mention—depends on continuity over the eons, something we don't see any evidence for in the archaeological record. What we do see, in contrast, is very suggestive evidence for the rather late and rapid emergence of symbolic thought, instead of the gradual long-term honing of the human capacity. So again, what might that almost instantaneous trigger have been? To us, it seems almost certain that it was the invention of language, combined with a brain and vocal tract that were already enabled for it. Language, after all, is the ultimate symbolic activity; and symbolic thought is almost inconceivable without it. Language involves forming and manipulating symbols in the mind, and recombining them in complete isolation from their referents in the outside world, to allow us to ask questions such as "What if?" It goes far beyond simple vocabulary; after all, chimpanzees can acquire vocabularies of symbols—even if they can't do

FIGURE 113. The rock shelters of Klasies River Mouth, at the base of the cliff in the center of this view, lie close to the southern tip of Africa. They were occupied by humans as much as 115 thousand years ago, and may contain the earliest evidence for a symbolic division of living space. Photograph by Willard Whitson.

it using speech, either because they don't have the right peripheral speech-producing equipment or because they lack the right connections in the brain.

Involving an infinitely flexible system of rules of syntax and grammar, language allows us to produce an infinite number of meanings from a finite number of elements. These rules and the ability to apply them are critical, for not all combinations will make sense. *Man paints house* would be a believable, if unlikely, headline to read, while you know that *house paints man* is an error. It is this structuring that makes human language so unique, and it is what makes it the ideal medium for the manipulation of mental symbols. Of course, once you have language not all of your thought processes need be mediated by symbols; the old intuitive substrate is still there. But while much that is creative in human thought derives from intuitive processes, the final output is linguistic, and symbolic.

We know that languages can be invented by human beings without too much difficulty, as witness the recent spontaneous emergence of a sign language among deaf Nicaraguan children. So in what context might language have been invented by nonlinguistic human beings with "symbol-ready" brains? People have had a lot of fun with this one, and the British anthropologist Robin Dunbar suggested not long ago that language arose as a sort of substitute for the physical grooming that primates of all sorts spend so much of their time doing to each other. Mutual grooming, which consists of picking through the fur of another individual with the fingers (or, in the case of lemurs, with specialized front teeth), is an intensely important social activity, cementing social bonds within the group. But it's not a very practical activity among largely hairless bipeds. Conversation—or, as Dunbar prefers to call it, gossip – does perform a very similar bonding function, which becomes more important as a cohesive influence the larger a group grows. What's more, Dunbar believes that there is a sort of feedback between language and social complexity, each driving the other. Still, this sort of scenario, attractive as it is, again envisages a long, gradual process, based on the underlying assumption that, feedback or not, evolution and optimization are effectively the same thing—which we know is not the case.

We prefer to think that, instead of having gradually evolved in a context of power plays and scheming, language might simply have been invented by children, in the context of play. Children are infinitely less hidebound than adults are, and less in thrall to established ways of doing things. When macaque monkeys in Japan invented the notion of washing the grit off sweet potatoes that were thrown to them on a sandy beach, it was the juveniles in the large group who took up that clearly advantageous innovation first. Later, the young adult females followed, then the older females, and only finally the males. Indeed, some of the crustier old males refused to wash the sweet potatoes at all, apparently preferring to damage their teeth on gritty tubers.

Language could easily have quite rapidly spread in a similar way through a population of symbol-ready but initially nonlinguistic *Homo sapiens*. With the cascading effect of symbolic thought having an impact on virtually every area of human experience and activity—song, dance, art, music, hunting, planning, you name it—it is not hard to see the advantage this innovation would have conferred on the group that thought it up. The subsequent spread of language among symbol-ready *Homo sapiens*, either by peaceful cultural contact or by conquest or more probably by both, would soon have given the entire species an advantage that it was evidently not at all reluctant to exercise–to the rapid detriment of all those other hominid species who then populated the Old World and, if a bit less rapidly, to that of virtually all other life forms except cockroaches and pigeons.

ONE IN A BILLION

BIODIVERSITY EXPERTS ESTIMATES THAT THERE ARE BETWEEN 2 MILLION AND 100 MILLION SPECIES ON OUR PLANET. WHEN WE CONSIDER THAT

the current biodiversity on the planet probably represents only 0.01 percent of all species that have ever existed, our species *Homo sapiens* is obviously only a drop in the biodiversity bucket. The position that our species occupies in the tree of life is neither inferior nor superior to that of any other; its evolution followed, and continues to follow, the same biological rules that all other species did during their time on Earth. In short, there was nothing special about how *Homo sapiens* came to be. At varying degrees of remoteness, we share common ancestry with all other living things, and as a result we carry the imprint of this history in our genomes, our behaviors and our appearances. Nonetheless, we happen to be distinctive in some very important ways, such as our ability to manipulate our surroundings, including the other organisms on this planet and even other members of our own species. These abilities are unique to us and appear to have been acquired very recently in evolutionary terms.

To demonstrate this, we can take a stroll along the "Walk through Time" in the American Museum of Natural History's Hayden Planetarium. This shows just how vast cosmic time is compared with that during which our kind has been "human," as we understand that word today. The walk itself is about 360 feet long, with each foot representing the passage of about 75 million years. The origin of life on our planet comes way beyond the halfway point on the walk. The origin of vertebrates occurs only about 5 feet from its end. The most primitive mammals evolved about 1 foot from the end, and primates a mere 6 inches away. The common ancestor of humans and their closest ape relatives first appeared about an inch from the end, while the first members of the genus *Homo* show up a microscopic $1/10$ of an inch before we reach today. Language first appeared about $1/100$th of an inch from the end, and all of written human history occurred in the last $1/1,000$th of an inch – much, much less than the width of a hair on your head!

On the larger scale of things, then, becoming human has happened extremely rapidly. But again, this is nothing special. The evolutionary histories of most creatures seem to have witnessed short bursts of innovation, separated by long expanses of business as usual. What *is* unusual, though, is the strong tendency within our species to come up with constant technological innovations in the period since humans adopted settled lifestyles.

It may have been many tens of thousands of years since we acquired the unique symbolic cognition that underwrites the capacity to do this; but since then, we have evidently been undertaking an unceasing exploration of the abilities that this new potential gives us. We thrive on exploration, and on the quest to understand ourselves and our surroundings; and there is no reason to believe that even greater acceleration in discovery and technology does not lie in our future. This aspect of ourselves is, in practical terms, probably the most important of all our uniquenesses, and it will almost certainly have an enormous impact on our future prospects. It therefore becomes very important for us to ask questions about where technology, human nature and our evolutionary trajectory might take us, and why.

Are There Any More Skeletons in the Closet?

First, given the evolutionary thrust of this book, let's ask what surprises may be in store for us in terms of our knowledge of our fossil record. It turns out that this is perhaps the most difficult of all such questions to answer, precisely because human evolution has not taken the form of a succession of links in a steadily modifying chain. We can't simply extrapolate new links to predict what we might find, because the signal arising from the human fossil record over its last several decades of enormous expansion is one of diversity. Human beings and their precursors have always, it seems, experimented with the evidently very many ways there are to be hominid. Many new finds, even if not quite all, come as complete surprises, as witness the merely 18,000-year-old but tiny-bodied and even tinier-brained *Homo floresiensis*, recently announced from an Indonesian island. Nobody would ever have predicted that such a hominid would ever be discovered, and after several years of fierce debate, there is still nothing like a consensus about how best to interpret it.

What we can predict, however, is that 40 years from now, the human fossil record will look very different from what we think we know today. When we became involved in evolutionary biology 30 or 40 years ago, the human fossil record was not only much smaller than it is now, but techniques for analyzing it were much more primitive, and the picture that most scholars derived from it was very different from the signal we are getting today. Science, as we have said, is a system of provisional knowledge; and it is inconceivable that several decades hence our understanding of the human past will not make our present viewpoints look quaint and naïve. As long as funding and human ambition continue, there will be myriad new hominid fossils and new ways of looking at them – and surprises aplenty in store!

Are There More Genomes in the Closet?

One of the important aspects of understanding our future as a species will lie in the understanding of our genomes—which is why, in earlier chapters, we have spent so much time in trying to make clear their workings. So where will our understanding of genes go in the future? Earlier in this book we asked, "whose genome was sequenced?" It turns out that it doesn't really matter, for a lot of reasons.

For one thing, we now have technologies for getting near-whole genome sequence information from any human we might want. This advance stems from knowing that not every nucleotide in our genomes is susceptible to change. In fact, 99.9 percent of all the nucleotides in our genomes are the same as those in any other randomly chosen human. This number is even smaller if we choose a relative; but, in general, it translates to about 30,000 nucleotide changes between you and the average other person. What this means in terms of the genes is that in order to get at the differences between any two humans we only need to look at 30,000 of the 3 billion nucleotides in the genome.

In turn, this is much easier if you know just where to look for those nucleotides; and a new scientific initiative to find them has already been started. It is called the International Halotype Map Project (HapMap). The idea behind it is to create ways to find the majority of the nucleotide positions in the genome that vary, and to create molecular biology shortcuts to tell what and where these changes are in different human beings. With this approach, it isn't necessary to sequence all 3 billion nucleotides in a person's genome to get at the important nucleotide changes. Actually, as we've seen, other advances are happening that might make obtaining sequences so easy that shortcuts might not be needed to make it possible for everyone to carry his genome on a card in his back pocket. At the current rate of progress in technology, a $1,000 genome will be possible within the next few years.

Such technical advances, whether they be HapMap or $1,000 genomes, or whatever, promise to make it possible to get a pretty precise view of nearly every human being's genome. Almost certainly, complete genomic information for large numbers of humans will soon be a reality. Whether this information will be equally distributed across nations, cultures and socioeconomic groups is, of course, another question, but in any event these advances in genomics will render irrelevant the earlier question of whose genome was sequenced. When everyone's genome can potentially be sequenced, this information can be individualized; and individualized genomes will take away many of the guessing games that medical scientists have had to play for the past century when dealing with genetic disorders.

With such information we can learn more about human traits; and when we know more about the genes involved in diseases, human health maintenance will be enhanced. Or so goes the current thinking among the clinically oriented folk. Other scientists are asking about non-disease traits. Will we ever be able to understand how we go from a genomic sequence to a complex behavior like musical ability or intelligence? We might, but then again, why would we want to know?

Finally, on the 150th anniversary of the discovery of the first Neanderthal fossil (August 1856), scientists at the Max Planck Institute in Germany announced an initiative to sequence the entire genome of *Homo neanderthalensis*. As we have mentioned earlier, extracting DNA and sequencing it from fossils is a difficult task. For the most part, only very small fragments of DNA can be isolated from fossils, and Neanderthal fossils are no different. In fact, the average piece of DNA from a Neanderthal fossil is only a bit over 100 bases. Considering that the whole Neanderthal genome should be about 3 billion bases, what this means is that more than 30 million puzzle pieces have to be generated and assembled into a complete genome. Using conventional methodology doing a project like sequencing the Neanderthal genome is impossible, or nearly so. But a new breakthrough in sequencing technology, called 454 sequencing, holds the key to deciphering the Neanderthal genome. This method sequences 100–base pair fragments by the basketful—over 40 million bases in a day. Conveniently, these 100 base fragments are just the right size for the Neanderthal. Some scientists are just darn lucky. The Neanderthal genome sequence will tell us much about the divergence of this enigmatic species from our own lineage, and about what makes us unique.

More Impact than a Meteorite

Questions of gene technology raise significant issues of bioethics. It is important that we realize that technology can be misused, and ethical considerations are at the heart of understanding when and how this might happen. Medical ethicists are vigorously debating many issues raised by genomics and the technology that flows from it. These issues have an impact on the daily lives of people and include questions of access to information, ownership of genetic information, availability of insurance, who gets to use the technology and many more. Think about the future and what it holds for you. What will your great-grandchildren be doing? Where will they be? Will they be smarter than you? Will they live longer than you? What about their athleticism or their ability to sing or to make art? As we have mentioned, some scientists think we will be able to alter the genes in our genomes to make improvements in all these areas. Many such alterations should, in principle, make us healthier and live longer. But some scientists think we might also be able to change our genes to make ourselves look better, or be stronger or smarter or faster, or any of a whole slew of things that will not necessarily make us any healthier. Should we let such gene changes be made? If we do, who should be able to use them? Who will make laws governing them? These are all ethical questions. All humans need to be aware of the changes that can be made, and of how the results of millions of individual decisions might affect the human population as a whole (just as a preference for boy children in some regions of China has led to a dearth of girls, with dire consequences predicted).

There are also issues that we need to consider with respect to the future evolution of humans. Two examples from the literature will help illustrate this problem. Martin Kreitman and his colleagues at the University of Chicago suggest that, even when we think we are doing a good deed like eliminating a disease, we might not have considered the evolutionary consequences of doing so. This example takes us back to sickle cell anemia, via the following interesting imagined scenario: About 100,000 years ago, some advanced and very benign beings visit Earth, and do a thorough genomic analysis of *Homo sapiens*, the most technologically savvy species on the planet. These advanced beings notice that there is a genetic disorder in *Homo sapiens* that occurs in very high frequency. It causes severe anemia in the humans with the genetic predisposition to it. The gene in question is the ß-globin gene, and the disease is sickle cell anemia; and to our advanced space travelers, correcting the disorder is a simple matter of inoculating the human populations with a genetically engineered virus that allows the incorporation of a normal gene for the blood anemia. They do so and then leave, and when they return 100,000 years later, what might they find? Quite possibly, the entire human species has gone extinct because of their benign but evolutionarily ignorant intervention.

Why? Remember that the sickle cell anemia gene also confers resistance to malaria to those who have one copy of the gene. So by eliminating the sickle cell gene variant, the space travelers have also eliminated the genetic ability of *Homo sapiens* to fight malarial infections, and the small and scattered human populations are driven to extinction by the malarial disease. Whether or not this is a realistic scenario, quite apart from its sci-fi aspects, the point is that, by not considering the evolutionary consequences of genetic change we might induce in *Homo sapiens*, we, too, might be like those well-intentioned space travelers.

Our second example comes from Lee Silver, a geneticist, who came up with the following story line: Genetic enhancement for resistance to disease is an easy thing to understand. But genetic enhancement for other things such as intelligence, good looks or athleticism is a bit more complex. Silver suggests that if genetic enhancement is available, there will be a market for it. But it might be expensive, so that perhaps only the very wealthy will be able to afford it. He calls these people the *genrich* (for gene enriched) and the *naturals*. The genrich have access to gene enhancement because of their high socioeconomic status, while the naturals have no access to the technology, because they are restricted by economics. Silver then plays the evolutionary tape forward, and suggests that the isolation of the genrich from the naturals will result in a speciation event. While this scenario is very futuristic—it might have been written for a Hollywood movie— aspects of Silver's explanation of the science and economics of genetic enhancement are not implausible.

Beam Me Up, Scotty

Other changes that some scientists foresee concern going into outer space. How this might affect the future of *Homo sapiens* is completely conjectural, but we can make some stabs at predicting it. Taking our earlier discussion of evolutionary processes as a starting point, it is evident that the conditions for true biological innovation in our species simply do not exist on the planet right now. This is because the conditions that facilitated human evolution—conditions such as small population size and vast migratory distances—have vanished, for the present, at least. People anywhere on the globe are no more than a few hours' travel away from people anywhere else. As few as 500 years ago, many human populations were isolated from one another even if they were physically not very far apart. Now people can move easily around the continents and potentially have children with virtually anyone. Isolation of small local populations—small enough to have the genetic instabilities that allow for both the incorporation of genetic novelties and for speciation—is a thing of the past. Today, as we saw with the web-like structure of relationships determined by genes on the X chromosome, genetic distances are being made smaller and smaller by the new demographic conditions.

Think about this, and its implications for how humans may evolve. Humans will always want to explore, and some futurologists find it not unreasonable to think that humans will start to explore outer space in the near future. The people who do so will carry out their missions in small populations carried in spaceships. Remember that small population size makes the population more conducive to genetic drift. As a result, the populations sent to explore Space might accrue more evolutionary change than the populations left behind on Earth. So what changes might happen to human beings in outer space? The theoretical possibilities are endless. Some genetic changes might arise that offer a selective advantage to our human space explorers, with the consequent fixation of relatively bizarre appearances, behaviors or physiological traits. Maybe only very short periods of time would need to elapse between the departure of space travelers and their complete divergence from their Earthly cousins, and new species of *Homo* would arise rapidly under the extreme conditions of isolation and drift. Still, because all of these rather improbable events would depend on isolation, these space travelers would be irrelevant to unfolding history back on Earth.

Diamond in the Rough

Some scientists, Jared Diamond among them, have discussed the potential impact of society as a whole on the future of human evolution. Diamond's approach has been to examine our history using scientific and technological tools, and to make observations about human society that have had an impact on the course of human change. He called his most successful book *Guns, Germs and Steel*, because it encompasses the impacts of war, disease and technology on human biological and cultural change. Without human developments involving all of these considerations, Diamond suggests, our current human condition might be very different. We must thus factor in the impact of human culture and society—and not least, its discovery of genomics—on our overall evolutionary trajectory.

What's more, in the past human evolution has been deeply influenced by external environmental changes, and our own impacts on global and local environments may feed back into that process. Over the past 250 years, industrialization and human population expansion have hugely changed environments all over the world. Coping with and mitigating the effects of such changes is becoming a mounting challenge to mankind—and an increasingly daunting one.

Back to the Brain

We are the one species on the planet with the capability of examining our past, using scientific and technological approaches. This capability stems ultimately from our brains and the unique way they are organized. As we saw earlier, we have learned much over the past century about how the brain works, and sophisticated tools already available, let alone tools as yet undreamed of, will inevitably lead to a more detailed and stunning understanding of what goes on inside the physical brain and its relationship to the mind. Using genomics, we will most likely figure out which genes are involved in making our brain human while a bonobo's develops into a bonobo brain. Using genomics and visualization techniques we will most likely be able to characterize the molecular and physiological parameters of many of our higher cognitive functions, like language, musical ability and even, possibly, consciousness itself. As Eric Kandel suggests, we are entering an age of a new science of mind; and one of the major results of this new science will be a greater understanding of our cognitive abilities.

Because increasing cognitive sophistication has seemingly consistently characterized the human line over the past 2 million years at least, is it reasonable to expect more of the same in the future? Intuitively, one might be tempted to answer in the affirmative. But if you were thinking in terms of the structure of the biological brain, think again; for we have already seen that the conditions needed for biological – genetic – innovation in the human population just don't exist right now. And, short of some (all-too-imaginable) cataclysmic disaster that fragments and drastically reduces the human population once more, they won't.

Still, that doesn't mean the game is over, for it is quite evident that the full potential provided by our extraordinary cognitive capacities has not yet been exploited. Even though the achievements of the Cro-Magnons were mind-boggling in cultural terms, their technologies were modest by modern standards. Subsequent technological evolution has proceeded at an ever-accelerating rate, especially since the Industrial Revolution; and now the rate of change is higher than ever. Modern technologies of genomic science, such as gene enhancement, stem cell research, genetic engineering and others, will almost assuredly have an impact on our biological and social futures; and in that sense, they are likely to lie at the center of much heated and ongoing debate within our society.

All of this ferment is part of our future. The excitement is not over; the pinnacle has not been reached. It is just the emphasis that has changed —from a process of biological innovation to the exploration of new ways to use the biology we have.

FURTHER READING

CHAPTER 1

Darwin, Charles. 1839. *The Voyage of the Beagle*. London: Henry Colburn. (Many reprint editions available.)

_____. 1859. *On the Origin of Species by Means of Natural Selection*. London: John Murray. (Many reprint editions available.)

Eldredge, Niles. 2005. *Darwin: Discovering the Tree of Life*. New York: W. W. Norton & Co.

Humes, Edward. 2007. *Monkey Girl: Evolution, Education, Religion, and the Battle for America's Soul*. New York: HarperCollins.

Popper, Karl R. 1959. *The Logic of Scientific Discovery*. New York: Basic Books.

Websites

Karl Popper: http://www.eeng.dcu.ie/~tkpw/

DarwinBlogs:

http://www.nileseldredge.com/darwin_blogs.htm

CHAPTER 2

Conroy, Glenn C. 1997. *Reconstructing Human Origins: A Modern Synthesis*. New York: W. W. Norton and Co..

Delson, Eric, Ian Tattersall, John Van Couvering, and Alison Brooks (eds.). 2000. *Encyclopedia of Human Evolution and Prehistory*. New York: Garland Publishing.

deWaal, Frans. 2005. *Our Inner Ape: A Leading Primatologist Explains Why We Are Who We Are*. New York: Riverhead Books.

Holloway, Ralph L., Douglas C. Broadfield, and Michaeol S. Yuan. 2004. *The Human Fossil Record, Vol 3: Brain Endocasts–The Paleoneurological Evidence*. New York: Wiley-Liss.

Johanson, Donald and Blake Edgar. 2006. *From Lucy to Language*, 2nd ed. New York: Simon and Schuster.

Jolly, Alison. 1999. *Lucy's Legacy: Sex and Intelligence in Human Evolution*. Cambridge, Massachusetts: Harvard Universty Press.

Klein, Richard. 1999. *The Human Career*, 2nd ed. Chicago: University of Chicago Press.

Schwartz, Jeffrey and Ian Tattersall. 2002. *The Human Fossil Record*, Vol. 1: *Terminology and Craniodental Morphology of Genus* Homo *(Europe)*. New York: Wiley-Liss.

_____. 2003. *The Human Fossil Record*, Vol. 2: *Craniodental Morphology of Genus* Homo *(Africa and Asia)*. New York: Wiley-Liss.

_____. 2005. *The Human Fossil Record*, Vol 4: *Craniodental Morphology of Early Hominids (Genera* Australopithecus, Paranthropus, Orrorin*) and Overview*. New York: Wiley-Liss.

Stringer, Chris and Peter Andrews. 2005. *The Complete World of Human Evolution*. London and New York: Thames and Hudson.

Tattersall, Ian. 1995. *The Fossil Trail: How We Know What We Think We Know About Human Evolution*. New York: Oxford University Press.

Tattersall, Ian and Jeffrey L. Schwartz. 2000. *Extinct Humans*. Boulder, Colorado: Westview Press.

Wood, Bernard. 2005. *Human Evolution: A Very Short Introduction*. Oxford and New York: Oxford University Press.

Zimmer, Carl. 2005. *Smithsonian Intimate Guide to Human Origins*. New York: HarperCollins.

Website

Anne and Bernard Hall of Human Origins:

http://www.amnh.org/exhibitions/permanent/humanorigins/

CHAPTER 3

DeSalle Rob and Michael Yudell (eds.). 2002. *Genomic Revolution: Unveiling the Unity of Life*. Washington, DC: National Academy Press.

_____ (eds.). 2004. *Welcome to the Genome: A User's Guide to your Genetic Past, Present, and Future*. Hoboken, New Jersey: John Wiley & Sons.

Lewontin, Richard. 2000. *The Triple Helix : Gene, Organism, and Environment*. Cambridge: Harvard University Press.

Pinker, Steven. 2002. *The Blank Slate*. New York: Viking.

Polcovar, Jane. 2006. *Rosalind Franklin and the Structure of Life*. Greensboro, North Carolina: Morgan Reynolds.

Ridley, Matt. 2000. *Genome: The Autobiography of a Species in 23 Chapters*. New York: HarperCollins.

Websites

Dolan DNA Learning Center:
http://www.dnalc.org/home.html

Genomic Revolution:
http://www.amnh.org/exhibitions/genomics/

CHAPTER 4

Carroll, Sean B. 2006. *The Making of the Fittest: DNA and the Ultimate Forensic Record of Evolution*. New York: W. W. Norton & Co.

Coyne, Jerry A. and H. Allen Orr. 2004. *Speciation*. Sunderland, Massachusetts: Sinauer Associates.

Eldredge, Niles. 1985. *Unfinished Synthesis: Biological Hierarchies and Modern Evolutionary Thought*. New York: Oxford University Press.

_____. 1995. *Reinventing Darwin: The Great Debate at the High Table of Evolutionary Theory*. Hoboken. New Jersey: John Wiley and Sons.

_____. 2005. *Darwin: Discovering the Tree of Life*. New York: W. W. Norton and Co.

_____ (ed.). 1992. *Systematics, Ecology and the Biodiversity Crisis*. New York: Columbia University Press.

Gould, Stephen Jay. 2002. *The Structure of Evolutionary Theory*. Cambridge, Massachusetts: Belknap Press.

Lambert, David M. and Hamish G. Spencer. 1995. *Speciation and the Recognition Concept: Theory and Application*. Baltimore, Maryland: Johns Hopkins University Press.

Mayr, Ernst. 2001. *What Evolution Is*. New York: Basic Books.

Palumbi, Stephen R. 2001. *Evolution Explosion: How Humans Cause Rapid Evolutionary Change*. New York: W. W. Norton & Co.

Tattersall, Ian. 2002. *The Monkey in the Mirror: Essays on the Science of What Makes Us Human*. New York: Harcourt Brace.

Templeton, Alan R. 1994. Biodiversity at the Molecular Genetic Level: Experiences from Disparate Macroorganisms. *Philosophical Transactions*: *Biological Sciences* 345(1311): 59-64.

Nicholas Wade. 2006. *Before the Dawn: Recovering the Lost History of Our Ancestors*. New York: Penguin.

Website

University of California at Berkeley, Understanding Evolution: http://evolution.berkeley.edu/

CHAPTER 5

Baum, David A., Stacey DeWitt Smith, and Samuel S. S. Donovan. 2005. The Tree-Thinking Challenge. *Science* 310(5750): 979 – 980.

Cracraft, Joel and Rodger W. Bybee (eds.). 2007. *Evolutionary Science and Society: Educating a New Generation*. Washington, DC: American Institute of Biological Sciences.

Cracraft, Joel and Michael J. Donoghue (eds.). 2004. *Assembling the Tree of Life*. Oxford and New York: Oxford University Press.

Jones, Steve. 1993. *The Language of Genes: Solving the Mysteries of Our Genetic Past, Present and Future*. New York: Doubleday.

Marks, Jonathan. 2002. *What It Means To Be 98% Chimpanzee: Apes, People and their Genes*. Berkeley: University of California Press.

Relethford, John. 2003. *Reflections of Our Past: How Human History is Revealed in Our Genes*. Cambridge, Massachusetts: Perseus Books.

Website

Tree of Life: http://www.tolweb.org/tree/

CHAPTER 6

Arsuaga, Juan Luis. 2002. *The Neanderthal's Necklace: In Search of the First Thinkers*. New York: Four Walls Eight Windows.

Asfaw, Berhane, W. Henry Gilbert, Yonas Bayene, W. K. Hart, Paul Renne, G. Wolde Gabriel, E. S. Vrba, and T. D. White. 2002. Remains of *Homo erectus* from Bouri, Middle Awash, Ethiopia. *Nature* 416: 317-320.

Bischoff, James L., D. D. Shamp, A. Aramburu, J. L. Arsuaga, E. Carbonell, and J. M. Bermudez de Castro. 2003. The Sima de los Huesos hominids date to beyond U/Th Equilibrium (>350 kyr) and perhaps to 400-500 kyr: New radiometric dates. *Journal of Archaeological Science* 30: 275-280.

Finlayson, Clive. 2004. *Neanderthals and Modern Humans: An Ecological and Evolutionary Perspective*. Cambridge, England: Cambridge University Press.

Gibbons, Ann. 2006. *The First Human: The Race to Discover Our Earliest Ancestors*. New York: Doubleday.

Hart, Donna and Robert W. Sussman. 2005. *Man the Hunted: Primates, Predators and Human Evolution*. New York: Westview Press.

Hublin, Jean- Jacques. 2001. Northwestern African Middle Pleistocene Hominids and their Bearing on the Emergence of *Homo sapiens*. In: Barham, L. and K. Robson-Brown (eds.), *Human Roots: Africa and Asia in the Middle Pleistocene*. Bristol: Western Academic and Specialist Press.

Johanson, Donald and Maitland Edey. 1981. *Lucy: The Beginnings of Humankind*. New York: Simon and Schuster.

Kalb, Jon. 2001. *Adventures in the Bone Trade: The Race to Discover Human Ancestors in Ethiopia's Afar Depression*. New York: Copernicus Books.

Kimbel, William H., Yoel Rak, and Donald C. Johanson. 2004. *The Skull of* Australopithecus afarensis. New York: Oxford University Press.

Kingdon, Jonathan. 2003. *Lowly Origin: Where, When and Why Our Ancestors First Stood Up*. Princeton, New Jersey: Princeton University Press.

Klein, Richard and Blake Edgar. 2002. *The Dawn of Human Culture*. Hoboken, New Jersey: John Wiley and Sons.

Krings, Matthias, Ann Stone, Ralf W. Schmitz, H. Krainitzki and Svante Pääbo. 1997. Neandertal DNA sequences and the origin of modern humans. *Cell* 90: 19-30.

Lewin, Roger. 1993. *The Origin of Modern Humans*. New York: Scientific American Library.

Mellars, Paul. 1996. *The Neanderthal Legacy: An Archaeological Perspective from Western Europe*. Princeton, New Jersey: Princeton University Press.

Schick, Kathy D. and Nicholas Toth. 1993. *Making Silent Stones Speak: Human Evolution and the Dawn of Technology*. New York: Simon and Schuster.

Shipman, Pat. 2001. *The Man Who Found the Missing Link: Eugene Dubois and His Lifelong Quest to Prove Darwin Right*. New York: Simon and Schuster.

Stanford, Craig B. 2003. *Upright: The Evolutionary Key to Becoming Human*. Wilmington, Massachusetts: Houghton Mifflin.

Stanford, Craig B. and Henry Bunn (eds). 2001. *Meat Eating and Human Evolution*. Oxford and New York: Oxford University Press.

Stanley, Steven M. 1996. *Children of the Ice Age: How a Global Catastrophe Allowed Humans to Evolve*. New York: Harmony Books.

Stringer, Chris and Clive Gamble. 1993. *In Search of the Neanderthals: Solving the Puzzle of Human Origins*. London and New York: Thames and Hudson.

Stringer, C. B., Jean-Jacques Hublin, and Bernard Vandermeersch. 1984. The origin of anatomically modern humans in western Europe. In: Fred Smith and Frank Spencer (eds), *The Origins of Modern Humans: A World Survey of the Fossil Evidence*. New York: Alan R. Liss.

Stringer, Chris and Robin McKie. 1996. *African Exodus: The Origins of Modern Humanity*. New York: Henry Holt.

Swisher, Carl C., Garniss H. Curtis, and Roger Lewin. 2000. *Java Man: How Two Geologists' Dramatic Discoveries Changed Our Understanding of the Evolutionary Path to Modern Humans*. New York: Scribner.

Tattersall, Ian. 1998. *The Last Neanderthal: The Rise, Success, and Mysterious Extinction of Our Closest Human Relatives*, revised ed. Boulder, Colorado: Westview Press.

van Andel, Tjeerd and William Davies. 2003. *Neanderthals and Modern Humans in the European Landscape During the Last Glaciation*. Cambridge, England: McDonald Institute.

Walker, Alan and Richard Leakey (eds.), 1993. *The Nariokotome* Homo erectus *Skeleton*. Cambridge, Massachusetts: Harvard University Press.

Walker, Alan and Pat Shipman. 1996. *The Wisdom of the Bones: In Search of Human Origins*. New York: Alfred A. Knopf.

Websites

American Museum of Natural History: http://www.amnh.org/exhibitions/permanent/humanorigins/

National Museum of Natural History, Smithsonian: http://www.mnh.si.edu/anthro/humanorigins/

CHAPTER 7

Cavalli-Sforza, Luigi Luca. 2000. *Genes, Peoples, and Languages*. New York: North Point Press.

Jablonski, Nina G. 2006. *Skin: A Natural History*. Berkeley: University of California Press.

Koppel, Tom. 2003. *Lost World: Rewriting Prehistory: How New Science is Tracing America's Ice Age Mariners*. New York: Atria Books.

Olson, Steve. 2002. *Mapping Human History: Discovering the Past Through Our Genes*. Wilmington, Massachusetts: Houghton Mifflin.

Sykes, Bryan. 2001. *The Seven Daughters of Eve: The Science That Revealed Our Genetic Ancestry*. New York: W. W. Norton & Co.

_____ (ed.). 1999. *The Human Inheritance: Genes, Language and Evolution*. Oxford: Oxford University Press.

Wade, Nicholas. 2006. *Before the Dawn: Recovering the Lost History of our Ancestors*. New York: Penguin.

Wells, Spencer. 2003. *The Journey of Man: A Genetic Odyssey*. Princeton, New Jersey: Princeton University Press.

Websites

Genographic: https://www3.nationalgeographic.com/genographic/

CHAPTER 8

Ackerman, Sandra. 1992. *Discovering the Brain*. Washington, DC: National Academy Press.

Allman, John M. 1999. *Evolving Brains*. New York: Scientific American Library.

Calvin, William H. 1996. *How Brains Think: Evolving Intelligence, Then and Now*. New York: Basic Books.

_____. 2002. *A Brain for All Seasons*. Chicago: Chicago University Press.

Changeux, Jean-Pierre. 1985. *Neuronal Man: The Biology of Mind*. Princeton, New Jersey: Princeton University Press.

Dobbs, David. 2005 Fact or Phrenology? The growing controversy over fMRI scans is forcing us to confront whether brain equals mind. *Scientific American Mind* April: 24-31.

Eccles, John C. 1989. *Evolution of the Brain: Creation of the Self*. New York: Routledge.

Jerison, Harry J. 1991. *Brain Size and the Evolution of Mind* (59th James Arthur Lecture). New York: American Museum of Natural History.

Kandel, Eric R. 2006. *In Search of Memory: The Emergence of a New Science of Mind*. New York: W. W. Norton & Co.

Kotulak, Ronald. 1996. *Inside the Brain: Revolutionary Discoveries of How the Mind Works*. Kansas City, Missouri: Andrews McNeel Publishing.

Parker, Sue Taylor and Michael L. McKinney. 1999. *Origins of Intelligence: The Evolution of Cognitive Development in Monkeys, Apes and Humans*. Baltimore, Maryland: Johns Hopkins University Press.

Ramachandran, V. S. and E. M. Hubbard. 2001. Synaesthesia: A window into perception, thought and language. *Journal of Consciousness Studies* 8(12): 3-34

Scientific American, (eds.). 1999. *The Scientific American Book of the Brain*. New York: Lyons Press.

Tattersall, Ian. 2002. *The Monkey in the Mirror: Essays on the Science of what Makes Us Human*. New York: Harcourt Brace.

Tomasello, Michael and Josep Call. 1997. *Primate Cognition*. New York: Oxford University Press.

Wade, Nicholas (ed.). 1998. *The Science Times Book of the Brain*. New York: Lyons Press.

Websites

Comparative Brain Collection: http://www.brainmuseum.org/

CHAPTER 9

Bickerton, Derek. 1995. *Language and Human Behavior*. Seattle: University of Washington Press.

Corballis, Michael C. 2002. *From Hand to Mouth: The Origins of Language*. Princeton, New Jersey: Princeton University Press.

Damasio, Antonio. 1999. *The Feeling of What Happens: Body and Emotion in the Making of Consciousness*. New York: Harcourt Brace.

Donald, Merlin. 2001. *A Mind So Rare: The Evolution of Human Consciousness*. New York: W. W. Norton & Co.

Gibson, Kathleen R. and Tim Ingold. 1993. *Tools, Language and Cognition in Human Evolution*. Cambridge, England: Cambridge University Press.

Jablonski, Nina and Leslie C. Aiello (eds.). 1998. *The Origin and Diversification of Language*. Mem. Cal. Acad. Sci. 24: 1-202.

Klein, Richard and Blake Edgar. 2002. *The Dawn of Human Culture*. Hoboken, New Jersey: John Wiley and Sons.

Lieberman, Philip. 2006. *Toward an Evolutionary Biology of Language*. Cambridge, Massachusetts: Belknap Press.

Macphail, Euan M. 1998. *The Evolution of Consciousness*. New York: Oxford University Press.

Marcus, Gary. 2004. *The Birth of the Mind: How a Tiny Number of Genes Creates the Complexities of Human Thought*. New York: Basic Books.

Mellars, Paul and Kathleen Gibson (eds.). 1996. *Modeling the Early Human Mind*. Cambridge, England: McDonald Institute.

Mithen, Steven. 2006. *The Singing Neanderthals: The Origins of Language, Mind, and Body*. Cambridge, Massachusetts: Harvard University Press.

Pinker, Steven. 1997. *How the Mind Works*. New York: W. W. Norton & Co.

Ruhlen, Merritt. 1994. *The Origin of Language: Tracing the Evolution of the Mother Tongue*. Hoboken, New Jersey: John Wiley & Sons.

Savage-Rumbaugh, Sue, Stuart G. Shanker, and Talbot J, Taylor. 1998. *Apes, Language and the Human Mind*. New York: Oxford University Press.

Tattersall, Ian. 1998. *Becoming Human: Evolution and Human Uniqueness*. New York: Harcourt Brace.

INDEX

opsin, 183-185
orangutan, 100, 108-110
Orrorin tugenensis, 113
Out-of-Africa hypothesis, 148

Pääbo, Svante, 110, 188
painting, 12, 189, 195-196
paleoanthropology, 11,21-23, 32-34, 35-36, 38, 41, 44-45, 69, 87, 89
Paleolithic, 44-45, 154-157, 197
Pan, 111
Pan troglodyles, 70
Panglossian Paradigm, 87, 89
parsimony, 94, 96
Patient 19, 78
pedigree, 70, 80, 143-146
phenome, 62
phenomics, 61-62
phenotype, 53, 61
Pingalap Island, 78-79, 183
Pithecanthropus erectus, 121
plasmid, 64
poly U, 75
polydactylism, 78
polymerase, 72-73
polyomavirus, 162
polymicrogyria, 187
Popper, Karl, 19-20, 42
population, 18, 23-25, 39-40, 49, 71, 71-86, 88-89, 115, 123, 127, 130, 136-137, 10-141, 143, 145-149, 151-152, 154-162, 181, 186, 189, 193, 201, 206-208; African, 161; American, 159; Amish, 78; Asian, 158, 162; Australian, 154; European, 136, 149, 155-156, 161; fish, 88; founder, 81, 158; genetics, 24, 71; *Homo erectus*, 193; Icelandic, 81; Polynesian, 161
porphyria, 78
Post-Glacial Expansion, 155
potto, 108
pre-Clovis, 159
primary microcephaly (MCPH), 187
primate, 24-46, 67, 82, 77, 107-108, 113-114, 139, 163-164, 167, 201, 203; brain, 187; genome, 24; higher, 199; hominid, 82; nonlinguistic, 199; Order, 46; social, 200; tree, 107
prokaryote, 93, 100-103
proline, 74, 76
protanomaly, 183-184
protanopia, 183-184
protein, 23, 49, 52, 56, 88, 66-63; 73, 76, 84, 119, 140, 147-148, 169-170, 179, 182, 184-186, 189
proteomic, 61-63
protostome, 103

quadriplegia, spastic, 78

quadruped, 115, 117; ancestral, 42, 114

random (genetic change), 15, 24, 64-65, 77-78, 82, 86, 88
recombination, 140, 164, 196
reductionism, 45, 82
replacement hypothesis, 135-136, 148, 154
reproductive success, 84-85, 199
ribosome, 52
RNA (ribonucleic acid), 49, 52, 62-63, 73-74, 76, 82, 161, 179-180, 187-188
Rowen, Lee, 60

Sacks, Oliver, 79
Sahelanthropus tchadensis, 113
Scopes trial, 25
sense, visual, 79, 172, 180-183, 185; olfactory, 170, 172, 180-181, 186; taste, 170, 176, 181; auditory, 109, 182, 174
sequence, 24, 29, 64, 66; DNA, 64, 96, 109, 205; genome, 24, 30, 49, 53, 64, 66, 96, 146, 205; mtDNA, 109
Severe Acute Respiratory Syndrome (SARS), 30
sex, 139-140, 168; change, 80; chromosome, 58, 139
sickle cell, 84, 86, 206; allele, 84-85; anemia, 84, 86, 206; disease, 85; resistance, 85
Silver, Lee, 107
SINE (short interspersed repetitive element), 60
Sinsheimer, Robert, 65
Sjogren, Torsten, 78
Sjogren-Larsson Syndrome, 78
sociogenomics, 61
speciation, 40, 69-71, 207
specific language impairment (SLI), 87
spiral cleavage, 103
Springsteen, Bruce, 35
Stone Age, 22; Middle, 131, 197; Old, 44
Sulston, John, 53
swan, black (*Cygnus atratus*), 19
symmetry, 93; bilateral, 103
syndrome, 78, 80, 84, 174, 183, 185-187; color-blind, 183, 185; developmental verbal dyspraxia (DVD), 185; Ellis-van Creveld, 78; Joubert (JBTS), 187; respiratory (SAR), 30; Sjogren-Larsson, 78; Williams, 162
systematics, 9, 11, 71, 89, 93-96, 100, 102, 105

T-cell, 162
tarsier, 46, 107-108
taxonomy, 93-93

Templeton, Alan, 148-149
thalassemia, 85
theory, 20-21, 26, 28-29, 112, 158-159; coalescent, 149, 158; Darwin's, 26, 29; evolutionary, 20-22, 25, 28-30; of bottleneck, 81; of Evolution, 26-28, 55, 192; of Mind, 199
thymine (T), 52, 54-55, 57-58, 65, 72-74, 95
tool, 45, 123, 125-127, 131-132, 179-182, 186, 191, 194, 196, 208; flint, 38; Oldowan, 119; pre-pared-core, 127; stone, 22, 44-45, 119-121; 123-124, 126-127, 189, 194, 198, 208; -maker, 119-120, 123, 126-127
trait, 52-53, 70, 86-88, 95, 168, 178-179, 182, 185, 205; anatomical, 178; complex, 86, 182; deleterious, 141; guevodoces, 88; reproductive, 86
transcription, 62-63, 186
transcriptomic, 61-63
tree (evolutionary), 18, 25, 42, 91-111, 113-115, 118, 136, 146, 153, 164; ape, 111; chromosomal, 144, 164; DNA, 164; family, 40; genomic, 42; mammal, 106; of Life, 46, 91, 93, 102-105, 107-109, 145, 170, 203; phylogenetic, 94-97, 165; primate, 107; systematic, 95
trichromacy, 184
tritanomaly, 183-184
tritanopia, 183-184
Turkana Boy, 121-123
typology, 45

Underhill, Peter A., 146

van Creveld, Simon, 78
variation (biological), 15-16, 18, 25, 40, 42, 55, 71, 82, 148, 165
Venter, Craig, 30, 66
vertebrate, 34, 93, 103-105, 126, 147, 171, 186, 197, 203; ancestral, 104
virus, 82-83, 161-163, 206
Voltaire (François-Marie Arouet), 86

Wallace, Alfred Russel, 17, 191-192
Watson, James, 57, 66, 71
Wexler, Nancy, 59
Wilkins, Maurice, 57
Williams Syndrome, 182
Wilson, Alan, 145, 147, 188
Woese, Carl, 93, 100-101

X chromosome, 58, 139, 141-143, 145, 163-165, 183-185, 207

Y chromosome, 136, 139-148, 151-154, 157-158, 160, 162, 164